T0296768

INTRODUCTION TO FRACTURE MECHANICS

INTRODUCTION TO FRACTURE MECHANICS

ROBERT O. RITCHIE
University of California Berkeley, United States

DONG LIU
University of Bristol, United Kingdom

ELSEVIER

Elsevier
Radarweg 29, PO Box 211, 1000 AE Amsterdam, Netherlands
The Boulevard, Langford Lane, Kidlington, Oxford OX5 1GB, United Kingdom
50 Hampshire Street, 5th Floor, Cambridge, MA 02139, United States

Notices
Knowledge and best practice in this field are constantly changing. As new research and experience broaden our understanding, changes in research methods, professional practices, or medical treatment may become necessary.

Practitioners and researchers must always rely on their own experience and knowledge in evaluating and using any information, methods, compounds, or experiments described herein. In using such information or methods they should be mindful of their own safety and the safety of others, including parties for whom they have a professional responsibility.

To the fullest extent of the law, neither the Publisher nor the authors, contributors, or editors, assume any liability for any injury and/or damage to persons or property as a matter of products liability, negligence or otherwise, or from any use or operation of any methods, products, instructions, or ideas contained in the material herein.

Library of Congress Cataloging-in-Publication Data
A catalog record for this book is available from the Library of Congress

British Library Cataloguing-in-Publication Data
A catalogue record for this book is available from the British Library

ISBN: 978-0-323-89822-5

For information on all Elsevier publications visit our
website at https://www.elsevier.com/books-and-journals

Publisher: Matthew Deans
Acquisitions Editor: Christina Gifford
Editorial Project Manager: Isabella C. Silva
Production Project Manager: Kamesh Ramajogi
Cover Designer: Greg Harris

Typeset by TNQ Technologies

Working together
to grow libraries in
developing countries

www.elsevier.com • www.bookaid.org

Contents

Preface

This book presents a somewhat personalized introduction to the origins, formulation, and application of fracture mechanics for the design, safe operation, and life prediction in structural materials and components. It is not our intent though to provide a formal treatment of the discipline as all of the mechanics derivations and formulations are readily available in the literature and in numerous text books. The intent here is to introduce and inform the reader how fracture mechanics works and how it is so different from other forms of analysis used to characterize mechanical properties. We therefore try to present the reader with the appreciation of these "difficulties" of fracture mechanics because it is our impression that the use of the discipline is often compromised by a lack of understanding.

We cover the topics of the foundation and use of linear-elastic fracture mechanics, involving both K-based characterizing parameter and G-based energy approaches, to characterize the fracture toughness of materials under plane-strain and non-plane-strain conditions, the latter using the notion of crack-resistance or R-curves. We follow that with a description of the far more complex nonlinear-elastic fracture mechanics based on the use of the J-integral and the crack-tip opening displacement. These topics largely involve continuum mechanics descriptions of crack initiation, slow crack growth, and eventual instability by overload fracture, but we attempt to couple this with mechanistic interpretations of the fracture modes using simple micromechanics formulations. Because of this, the reader may note that we do present somewhat of a bias for the characterizing parameter approaches, rather than the energy approaches, even though they are inseparable because the former descriptions are more amenable to local micromechanical modeling. We conclude with a description of the application of fracture mechanics to subcritical crack growth, specifically by environmentally-assisted cracking, creep-crack growth and especially fatigue, followed by a final chapter on worked examples of fracture mechanics in practice, with examples involving failure by plastic yielding *vs.* fracture, the leak-before-break concept applied to pressure vessels, the fracture of pre-tensioned bolts, and estimating the safe lifetime of medical heart valve prostheses. The reader may be disappointed that we do not cover such topics as dynamic fracture and interfacial failure or treat in detail the application of fracture mechanics to all forms of subcritical cracking, but in the interests of keeping this introductory treatment relatively succinct and concise, we only provide a brief description of some of these more advanced topics.

This introduction is intended for anyone interested in the field of fracture, and particularly for students, researchers, professors, and practicing engineers who need to get started in this discipline and to understand what is behind this somewhat strange form of mechanics. Indeed, among purists, the topic has often aroused suspicion. When fracture mechanics initially arrived on the scene in the late 1950s to 1960s, many physicists could not comprehend how its fundamental parameter, at that time the stress-intensity factor K, could have such weird units as ksi$\sqrt{\text{in}}$, *i.e.*, with dimensions of stress times the *square-root* of distance!

One cannot underestimate the impact of fracture though. Consider the loss of the RMS Titanic in 1912, the failure of the DeHavilland Comet commercial jet airliners in the 1950s, and Lockheed DC-10 aircraft some 20 to 30 years later. These were truly traumatic events that in many respects "changed the world". As someone once said: "Whereas God may have invented plasticity, the Devil invented fracture!" Need we say more?

CHAPTER

1

Introduction

Fracture mechanics was developed comparatively recently, essentially after the end of the Second World War, to furnish an engineering approach that could be employed to quantitatively assess the onset of fracture. As such, it has played a vital role in enabling the creation of safer engineering structures, including those used in transportation (e.g., airframes, gas-turbine engines), the construction industries (e.g., supporting beams, welded structures), and energy production (e.g., power turbines, pressure vessels, and piping), in establishing (nominal) material properties to measure the fracture toughness for the development and characterization of structural materials, and most importantly in linking these length-scales together. In principle, for a given material which could be a metal, ceramic, or polymer, a fracture toughness value can be measured on a relatively small sample in the laboratory and then used directly to predict the fracture of a much larger structure or component in service. Similar analyses can be performed for failures by fatigue, creep, or environmentally-assisted induced fracture. Powerful? In engineering terms, most certainly. However, when one looks at the literature in this field, one quickly comes to the conclusion that fracture mechanics has become one of the most "abused" (for want of a better word) form of mechanics, with, for example, many measured fracture toughness values quoted in technical papers being at worst "dead wrong" or at best ge-ometry and specimen-size dependent.

In light of this quandary, in this primer we will present the rudiments of fracture mechanics, not specifically in terms of its fundamental origins based on complex applied and computational mechanics, as this has been described numerous times in various textbooks, but rather from the perspective of the philosophy underlying the few principles and, yes, the assumptions that form the basis of the discipline. We will attempt to provide the reader with a "working knowledge" of fracture mechanics, to describe its potency for damage-tolerant design, for preventing failures through appropriate life-prediction strategies, and for quantitative failure analysis (fracture diagnostics), while at the same time communicating a necessary understanding of the methodology to avoid the many "pitfalls" that seem to be implicit with its use. In simple terms, this primarily means being cognizant of the many engineering approximations and assumptions that need to be made to afford a meaningful description of the stress and displacement fields in the vicinity of a crack, and how this can be utilized to specify parameters, such as the stress-intensity factor K and strain-energy release rate G for linear-elastic materials or the J-integral for nonlinear-elastic materials, that can be measured to characterize the initiation of cracking and in certain cases its subsequent stable propagation, i.e., to define a "crack-driving force."

To quote the words of Dan Drucker and Jim Rice: *"Fracture mechanics is the judicious interpretation of crack tip singular fields."* To our minds, a true understanding of this statement represents a necessary appreciation of what fracture mechanics is and how it can be faithfully applied realistically to prevent engineering failures.

2

Foundations of fracture mechanics

2.1 Ideal fracture strength

One place to start a description of fracture mechanics is to consider the atomistic ideal fracture (or "cohesive") strength. In essence, this, in its simplest form, represents the stress required to pull two planes of atoms, e.g., cleavage planes, apart. It is known that two interacting atoms are subjected to two types of forces: a repulsive force at short ranges due to Pauli's exclusion principle, and a force of attraction (dispersion force) at long ranges. In 1931, John Lennard-Jones [1] at the University of Bristol estimated the potential energy of two atoms as a function of their separation distance, shown schematically in Fig. 2.1a and b. The atoms are at equilibrium when the potential energy is at a minimum, marked by a separation of l_0. At this point, the atoms are most stable and remain at this distance until an external force is exerted upon them. Assuming a far-field stress, σ^∞, is applied normal to the cleavage planes along the axis of the atomic bonds over a unit area, it can be approximated to follow a sinusoidal form with the distance, Δl:

$$\sigma^\infty = \sigma_T \sin\frac{\pi\Delta l}{\lambda},\tag{2.1}$$

where λ is of a length comparable to l_0, and σ_T is the amplitude of the sinusoidal stress, i.e., the maximum cohesive stress (Fig. 2.1c).

For a linear-elastic solid, the stress applied to increase the atoms' separation by Δl from the equilibrium spacing l_0 can be expressed in terms of Hooke's law:

$$\sigma^\infty = E\Delta l/l_0,\tag{2.2}$$

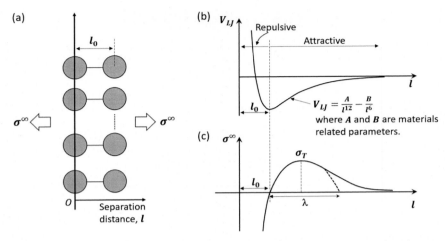

FIGURE 2.1 (a) A schematic of two planes of atoms at equilibrium distance l_0, subject to far-field stress, σ^∞; (b) the Lennard-Jones potential V_{LJ} curve, including a steep (short-range) repulsive term and a smoother (longer-range) attractive term; l_0 is the distance at which two simple atoms are at equilibrium; (c) the stress required to pull two atoms apart at equilibrium is simplified to a sinusoidal form with wavelength of 2λ; the maximum stress σ_T is the theoretical (ideal) cohesive fracture strength.

where E is the elastic modulus. Equating Eqs. (2.1) and (2.2), it is apparent that:

$$\sigma_T \sin\frac{\pi\Delta l}{\lambda} = \frac{E\Delta l}{l_0}. \tag{2.3}$$

At small angles, $\sin\frac{\pi\Delta l}{\lambda} \to \frac{\pi\Delta l}{\lambda}$, and as $\lambda \sim l_0$, we can obtain an estimate of the ideal fracture strength:

$$\sigma_T \sim \frac{E}{\pi}. \tag{2.4}$$

These calculations can be performed with far greater accuracy. For instance, more accurate descriptions of σ^∞ can replace the simplified sinusoidal form used here, yet all these solutions give the ideal fracture strength σ_T to be typically in the range of $E/4$ to $E/15$, with the customary value of σ_T being $E/10$. However, all these theoretical estimates are very high compared to the measured fracture strengths. For example, silica glass has a modulus of 70 GPa, which would give a theoretical strength on the order of 7 GPa. However, the experimentally measured tensile strength of glass is usually in the range of 30 to 100 MPa, which is more than an order of magnitude less. Why, you may ask? Well, analogous to the fact that the measured shear strengths of materials are typically two orders of magnitude smaller than the ideal shear strength due to the presence of dislocations, in the case of the fracture strength, this discrepancy between theoretical and experimental strengths is due to the inevitable presence of cracks, which results in local stress concentrations.

2.2 Griffith fracture theory

Since the (experimental) fracture strength involves the separation of atomic planes in the presence of such stress concentrations associated with pre-existing flaws and cracks, this can make any estimate of the fracture strength difficult because the stress concentration can be a function of the shape and size of the crack. The first recognizable theory to estimate the actual fracture strength, due to Griffith in the 1920s, used energy methods to avoid this problem [2]. He considered purely elastic conditions with an internal crack of length $2a$, in an infinite sheet of unit thickness subjected to an applied (far-field) tensile stress, σ^∞ (Fig. 2.2).

To estimate the stress to propagate this crack, Griffith considered the total energy, U_{Total}, of this system to be equated to the potential energy, U_{PE}, and the work to create two new fresh surfaces, W_S, as the crack extends:

$$U_{Total} = U_{PE} + W_s, \qquad (2.5a)$$

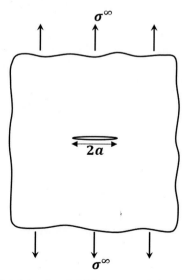

FIGURE 2.2 An infinite plate of unit thickness containing a middle crack of length $2a$ subjected to a far-field applied stress σ^∞.

where the potential energy is the elastic stored energy, U_ε, minus the work done, W_D:

$$U_{PE} = U_\varepsilon - W_D. \tag{2.5b}$$

By considering fixed displacement conditions, such that $W_D = 0$, the potential energy will equal the strain energy, which can be estimated as $U_\varepsilon = -\sigma^2 (\pi a^2)/E$, where E is Young's modulus. The corresponding energy to create new fresh surfaces would be the length of the crack, $2a$, times two (as there are two crack surfaces), multiplied by the surface energy per unit area of the fracture plane, γ_s, i.e., $W_s \sim 4a\gamma_s$, where γ_s is typically on the order of 1 J/m^2 (this is the actual value for silicon). The driving force for fracture is the release of strain energy which must be no less than the energy needed to create the fresh crack surfaces. This can be obtained by taking the derivative with respect to a of the total energy, $U_{Total} = 4a\gamma_S - \sigma^2(\pi a^2)/E$, and setting this equal to zero, such that unstable fracture will occur when

$$\sigma^\infty = \sigma_F = \sqrt{\frac{2\gamma_S E}{\pi a_c}}, \tag{2.6}$$

where σ^∞ is the applied (far-field) stress. This is Griffith's famous estimate of the actual fracture strength, σ_F, under ideally elastic conditions, which is notable because it depends not simply on the applied stress but also the presence and size of pre-existing cracks. The critical crack size, a_c, for fracture is thus the crack size above which the crack can extend with decreasing energy conditions (Fig. 2.3).

As this is an elastic theory, it cannot take into account the work required for plastic deformation (the plastic work term) which in the case of ductile materials can involve an energy orders of magnitude greater than γ_s. Accordingly, for such materials, the Griffith equation can severely underestimate the fracture strength. However, we can consider it as *a necessary but insufficient criterion for fracture*.

Irwin [3] and Orowan [4] later revised the Griffith equation (Eq. 2.6) to include the plastic work term, γ_p, which gives the expression:

$$\sigma_F = \sqrt{\frac{2(\gamma_s + \gamma_p)E}{\pi a_c}}, \tag{2.7}$$

but as $\gamma_p \gg \gamma_s$ for all ductile materials, it seems somewhat futile to use this equation containing a large plastic work term to predict the fracture strength when the relationship is based on ideal elasticity.

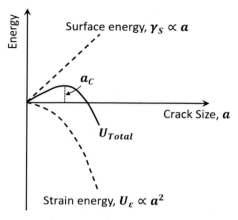

FIGURE 2.3 In the Griffith theory, the total energy, U_{Total}, for fracture in a center-cracked infinite plate with unit thickness, presents a competition between the strain energy U_ε that is released by fracture (the driving force) balanced by the energy required to create new crack surfaces. U_{Total} thus varies with crack size, a. Above a critical crack size, a_c, fracture is spontaneous, i.e., occurring without any further increase in total energy.

2.3 Orowan approach

An alternative (ideally elastic) approach to estimating the actual fracture strength, which in some respects is more insightful than Griffith, is the approach of Orowan which was proposed later [5]. Again, considering an internal crack of length $2a$, in an infinite sheet of unit thickness subjected to an applied tensile stress, σ^∞, the local stress at the tip of any crack can be estimated as $k_t \sigma^\infty$ where k_t is the stress concentration factor ($\sim 2\sqrt{a/\rho}$, where ρ is the root radius of the tip of the crack). Orowan then equated this local stress to the ideal cohesive strength at fracture, σ_T, which led to the expression:

$$\sigma^\infty = \sigma_F = \frac{\sigma_T}{2}\sqrt{\frac{\rho}{a}}. \tag{2.8}$$

In glass, for example, where the crack sizes are the order of micrometers whereas the root radius of any pre-existing cracks would be of order of an atomic spacing, i.e., in the nanometer range, this theory also predicts actual fracture strengths to be an order of magnitude or so lower than the ideal cohesive strength. These predictions are realistic and, as such, the Orowan model is certainly physically conceptual, but the approach could never be used in an engineering predictive manner without prior knowledge of the orientation, geometry, and root radius of the worst-case defects pre-existing in the solid.

2.4 Origins of fracture mechanics theory

The early developments to create a quantitative methodology to predict the onset of fracture were similarly based on ideal elasticity — linear-elastic fracture mechanics (LEFM) — which came basically from two sources: (i) an energy approach proposed by Irwin [6] which essentially is a more general version of the Griffith model [2] but now based on the strain-energy release rate G as the crack-driving force, where G is defined as the rate of change in potential energy per unit increase in crack area, and (ii) a parallel approach based on so-called *characterizing (or governing) parameters*, such as the stress-intensity factor K, which are *global* parameters that serve to "characterize" the *local* stress and displacement fields in the vicinity of a crack tip [7–10].

Many researchers often consider these two approaches to be somewhat in conflict, yet they are simply "parallel universes." For example, many in the polymer community seem to prefer the use of G, whereas the metal and ceramics communities invariably use the K approach. Although the use of G makes it easier to analyze the growth of cracks under mixed-mode conditions (e.g., tension plus shear), it is our belief that fracture mechanics based on the use of characterizing parameters is a more "honest" (for want of a better word) methodology; moreover, it is more amenable to coupling with the mechanistic aspects of fracture. Accordingly, we will begin with a description of linear-elastic fracture mechanics based on K as the characterizing parameter, and then subsequently describe the G-based approach (even though the latter more naturally follows from Griffith).

References

[1] J.E. Lennard-Jones, Cohesion, Proc. Phys. Soc 43 (1931) 461.
[2] A.A. Griffith, The phenomena of rupture and flow in solids, Philos. Trans. Roy. Soc 221 (1920) 163.
[3] G.R. Irwin, Fracture dynamics, in: Fracture in Metals, ASM, Cleveland, OH, 1948, p. 147.
[4] E. Orowan, Notch brittleness and the strength of metals, Trans. Inst. Engrs. Shipbuilders Scotland 89 (1945) 165.
[5] E. Orowan, Fracture and strength in solids, Rep. Prog. Phys. 12 (1948-49) 185.
[6] G.R. Irwin, Onset of fast crack propagation in high strength steel and aluminum alloys, Proc. Sagamore Res. Conf. 2 (1956) 289.
[7] H.M. Westergaard, Bearing pressures and cracks, J. Appl. Mech. 6 (1939) 49.

[8] I.N. Sneddon, The distribution of stress in the neighbourhood of a crack in an elastic solid, Proc. Roy. Soc. (London) A-187 (1946) 229.

[9] M.L. Williams, On the stress distribution at the base of a stationary crack, J. Appl. Mech. 24 (1957) 109.

[10] G.R. Irwin, Analysis of stresses and strains near the end of a crack traversing a plate, J. Appl. Mech. 24 (1957) 361.

3

Linear-elastic fracture mechanics (LEFM)

3.1 Stress analysis of cracks: Williams $1/\sqrt{r}$ singularity and stress-intensity factor K

3.1.1 Crack-tip fields

LEFM and the use of the *stress-intensity factor*[1] K originate from the analysis of cracks to define the local stress and displacement fields ahead of a sharp crack. In the years preceding the formal development of fracture mechanics, several solutions were derived for cracks in isotropic purely elastic bodies in static equilibrium. Of note here are the singular solutions of Westergaard [1], Sneddon [2], and Williams [3], which predict that the crack-tip stresses will decay as $1/\sqrt{r}$, where r is the distance ahead of the crack tip. Using the nomenclature defined in Fig. 3.1 in Box 3.1, in terms of Cartesian coordinates, these solutions can be written, as $r \to 0$ in the form:

$$\sigma_{ij} \to \frac{K}{\sqrt{2\pi r}} f_{ij}(\theta), \tag{3.1a}$$

$$u_i \to \frac{K}{2E} \sqrt{\frac{r}{2\pi}} f_i(\theta), \tag{3.1b}$$

where f_{ij} and f_i are dimensionless functions of θ, the inclination with respect to the crack plane ($\theta = 0$ directly ahead of the crack tip), as defined in Fig. 3.1, and E is Young's modulus. K is simply a singular characterizing parameter, unset by the field, but which uniquely characterizes the

---

[1] The *stress-intensity factor K*, or simply the *stress intensity*, with units of $F \cdot L^{-3/2}$ (F is force and L is length), must be distinguished from the *stress concentration factor*, which is the dimensionless ratio of maximum stress at a notch divided by the nominal stress.

11

BOX 3.1

Crack-tip stress and displacement fields for linear-elastic isotropic materials

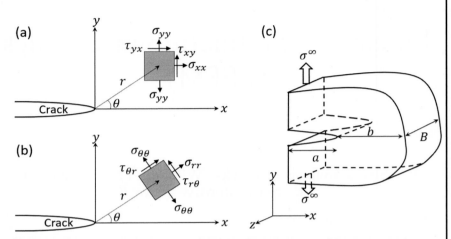

FIGURE 3.1   Stress nomenclature definition ahead of a crack tip in terms of (a) Cartesian and (b) polar coordinates; (c) specimen dimensions of crack length a, remaining uncracked ligament b, and thickness B.

For modes I, II, and III:

Local stresses:

$$\sigma_{ij} = \frac{K_i}{\sqrt{2\pi r}} f_{ij}(\theta) + HOT$$

Local displacement:

$$u_i = \frac{K_i}{2E}\left(\frac{r}{2\pi}\right)^{\frac{1}{2}} f_i(\theta) + HOT$$

where E is Young's modulus, r is distance ahead of the crack tip, $f_{ij}(\theta)$ and $f_i(\theta)$ are functions of angle θ around the crack tip, K_i is the stress-intensity factor in mode $i = 1$ to 3, and HOT refers to higher order terms.

BOX 3.1 *(cont'd)*

Crack-tip stress and displacement fields for linear-elastic isotropic materials

For mode I (tensile opening):

$$\sigma_{zz} = \nu(\sigma_{xx} + \sigma_{yy}) = \nu(\sigma_{rr} + \sigma_{\theta\theta}) \quad \text{[Plane strain]}$$
$$= 0 \quad \text{[Plane stress]}$$
$$\sigma_{xz} = \sigma_{yz} = 0 = \sigma_{rz} = \sigma_{\theta z}$$

$$\begin{pmatrix} \sigma_{xx} \\ \sigma_{yy} \\ \sigma_{xy} \end{pmatrix} = \frac{K_{\mathrm{I}}}{\sqrt{2\pi r}} \begin{pmatrix} -\cos\left(\dfrac{\theta}{2}\right)\left[1 - \sin\left(\dfrac{\theta}{2}\right)\sin\left(\dfrac{3\theta}{2}\right)\right] \\ \cos\left(\dfrac{\theta}{2}\right)\left[1 + \sin\left(\dfrac{\theta}{2}\right)\sin\left(\dfrac{3\theta}{2}\right)\right] \\ \sin\left(\dfrac{\theta}{2}\right)\cos\left(\dfrac{\theta}{2}\right)\cos\left(\dfrac{3\theta}{2}\right) \end{pmatrix}$$

$$\begin{pmatrix} \sigma_{rr} \\ \sigma_{\theta\theta} \\ \sigma_{r\theta} \end{pmatrix} = \frac{K_{\mathrm{I}}}{\sqrt{2\pi r}} \begin{pmatrix} \cos\left(\dfrac{\theta}{2}\right)\left[1 + \sin^2\left(\dfrac{\theta}{2}\right)\right] \\ \cos^3\left(\dfrac{\theta}{2}\right) \\ \sin\left(\dfrac{\theta}{2}\right)\cos^2\left(\dfrac{\theta}{2}\right) \end{pmatrix}$$

$$\begin{pmatrix} u_x \\ u_y \end{pmatrix} = \frac{K_{\mathrm{I}}}{2E}\left(\frac{r}{2\pi}\right)^{\frac{1}{2}} \begin{pmatrix} (1+\nu)\left[(2k-1)\cos\left(\dfrac{\theta}{2}\right) - \cos\left(\dfrac{3\theta}{2}\right)\right] \\ (1+\nu)\left[(2k-1)\sin\left(\dfrac{\theta}{2}\right) - \sin\left(\dfrac{3\theta}{2}\right)\right] \end{pmatrix}$$

$$\begin{pmatrix} u_r \\ u_\theta \end{pmatrix} = \frac{K_{\mathrm{I}}}{2E}\left(\frac{r}{2\pi}\right)^{\frac{1}{2}} \begin{pmatrix} (1+\nu)\left[(2k-1)\cos\left(\dfrac{\theta}{2}\right) - \cos\left(\dfrac{3\theta}{2}\right)\right] \\ -(1+\nu)\left[(2k-1)\sin\left(\dfrac{\theta}{2}\right) + \sin\left(\dfrac{3\theta}{2}\right)\right] \end{pmatrix}$$

where ν is Poisson's ratio, and $k = (3 - 4\nu)$ in plane strain and $\frac{3-\nu}{1+\nu}$ in plane stress.

continued

BOX 3.1   *(cont'd)*

Crack-tip stress and displacement fields for linear-elastic isotropic materials

For mode II (shear):

$$\sigma_{zz} = \nu(\sigma_{xx} + \sigma_{yy}) \quad \text{[Plane strain]}$$
$$= 0 \quad \text{[Plane stress]}$$
$$\sigma_{xz} = \sigma_{yz} = 0$$

$$\begin{pmatrix} \sigma_{xx} \\ \sigma_{yy} \\ \sigma_{zz} \end{pmatrix} = \frac{K_{\text{II}}}{\sqrt{2\pi r}} \begin{pmatrix} -\sin\left(\frac{\theta}{2}\right)\left[2 + \cos\left(\frac{\theta}{2}\right)\cos\left(\frac{3\theta}{2}\right)\right] \\ \sin\left(\frac{\theta}{2}\right)\cos\left(\frac{\theta}{2}\right)\cos\left(\frac{3\theta}{2}\right) \\ \cos\left(\frac{\theta}{2}\right)\left[1 - \sin\left(\frac{\theta}{2}\right)\sin\left(\frac{3\theta}{2}\right)\right] \end{pmatrix}$$

$$\begin{pmatrix} u_x \\ u_y \end{pmatrix} = \frac{K_{\text{II}}}{2E}\left(\frac{r}{2\pi}\right)^{\frac{1}{2}} \begin{pmatrix} (1+\nu)\left[(2k+3)\sin\left(\frac{\theta}{2}\right) - \sin\left(\frac{3\theta}{2}\right)\right] \\ -(1+\nu)\left[(2k+3)\cos\left(\frac{\theta}{2}\right) + \cos\left(\frac{3\theta}{2}\right)\right] \end{pmatrix}$$

For mode III (anti-plane shear):

$$\begin{pmatrix} \sigma_{xz} \\ \sigma_{yz} \end{pmatrix} = \frac{K_{\text{III}}}{\sqrt{2\pi r}} \begin{pmatrix} -\sin\left(\frac{\theta}{2}\right) \\ \cos\left(\frac{\theta}{2}\right) \end{pmatrix} \qquad \sigma_{xx} = \sigma_{yy} = \sigma_{xy} = 0$$

$$u_z = \frac{4K_{\text{III}}}{E}\left(\frac{r}{2\pi}\right)^{\frac{1}{2}}\left[(1+\nu)\sin\left(\frac{\theta}{2}\right)\right] \qquad u_x = u_y = 0$$

magnitude of the local stresses and displacements. Note that the full expressions in Eq. (3.1a,b) contain higher order terms which invariably (but not always) we can ignore as they are small enough, to be able to claim that a single parameter K fully characterizes the crack-tip stress and displacement fields. It is this K that we consider to be the singular governing parameter of the crack-tip field which has been identified as the so-called stress-intensity factor. To be consistent with the Griffith approach, it is now defined in terms of the applied (far-field) stress σ^∞ and crack size a, as:

$$K_i = Y\sigma^\infty \sqrt{\pi a}, \tag{3.2}$$

where Y is a geometry factor, which is often a function of the ratio of crack size to sample width (a/W), and $i = 1$ to 3 depending upon the crack displacements or modes (Fig. 3.2). The unit used for K is usually MPa√m; the corresponding English unit is ksi√in, where 1 ksi√in $\equiv$ 1.099 MPa√m.

Details of these crack-tip singular fields which are characterized by K are given in Box 3.1. The most common use of K is in mode I for the tensile opening of a crack. This is generally (but not always) the worst-case scenario as the fracture resistance is often lowest in this mode, but the approach can also be used for cracks in shear (mode II) where the characterizing parameter is K_{II}, and in anti-plane shear (mode III) where the characterizing parameter is K_{III}. Note that the value of K_i is determined by global (external) parameters which can be readily measured, such as the geometry, crack size, and applied stresses; yet it uniquely controls the magnitude and distribution of the local stresses and strains at the crack tip which are the ones that actually control fracture.

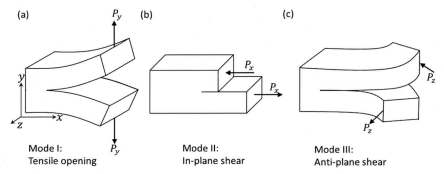

(a) Mode I: Tensile opening

(b) Mode II: In-plane shear

(c) Mode III: Anti-plane shear

FIGURE 3.2 Schematic of the three fundamental modes of loading of a crack: (a) tensile opening mode I, (b) in-plane shear mode II, and (c) anti-plane or out-of-plane shear mode III (the deformations of the crack faces in the schematic are exaggerated).

The geometry factors, Y, in Eq. (3.2), the so-called K-solutions, have mostly been derived numerically and are available in several handbooks [4,5]. Several solutions are shown in Fig. 3.3 in Box 3.2.[2] Moreover, as they are determined for linear-elastic constitutive behavior, the *principle of superposition* can be applied to calculate the value of K for a more complex loading situation involving more than one set of tractions. However, whereas you can superimpose two stress intensities of the same mode, you cannot superpose mixed-mode Ks, e.g., a K_I with a K_{III}. This is one disadvantage of the characterizing parameter approach, as it cannot directly address such mixed-mode loading situations which are not that uncommon in service. For these situations, the energy-based G approach is more amenable to mixed-mode loading, as described below in Section 3.6. However, at this stage, most of the following discussion focuses on the all-important mode I failures.

There are several issues here that are especially noteworthy. Firstly, these solutions (in Eq. 3.1a,b) for the elastic crack-tip stresses represent singularities, in that the tensile opening stress becomes infinite at the crack tip, i.e., $\sigma_{yy} \to \infty$ as $r \to 0$. Secondly, the actual solutions are characterized by a series of terms, but by considering these solutions as $r \to 0$, we are generally able to ignore (but again not always) all the higher order terms, such that the first term, controlled by a single characterizing parameter (in this case K_i), is dominant.[3] Thirdly, these solutions are planar fields, in the x-y plane in Fig. 3.1; they define the elastic distributions of the crack-tip σ_{yy}, σ_{xx}, and σ_{xy} stresses but say nothing directly about the out-of-plane stresses, e.g., σ_{zz}, which can exist through the thickness of the sample. These through-thickness stresses are set by the geometry and the deformation conditions, the extremes of which are plane stress vs. plane strain. This is described in more detail below and in Section 3.4 and Box 3.3.

However, for a given state of constitutive behavior, i.e., in the present case linear-elastic deformation for LEFM,[4] such K-fields are unique and autonomous; the distributions of local elastic stresses, strains, and

---

[2] Note that because of symmetry considerations, in these K-solutions, internal cracks are of length $2a$ whereas surface crack are of length a. This means that surface cracks of a given length are at least twice as potent as internal cracks of the same size.

[3] The irony of this is that one might say that these singular solutions are most accurate where they are least relevant!

[4] Although the overall (*global*) deformation conditions are linear elastic in LEFM, it is understood that some degree of plasticity must exist *locally*, e.g., at the tips of cracks. As noted below, the linear-elastic analysis can still be appropriate as long as the scale of this local plasticity is small enough to ignore.

BOX 3.2

Stress-intensity K_I solutions

Crack in infinite body

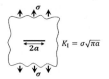

$K_I = \sigma\sqrt{\pi a}$

Edge crack in a semi-infinite body

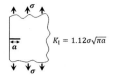

$K_I = 1.12\sigma\sqrt{\pi a}$

Edge crack in a semi-infinite body subjected to linearly varying stress

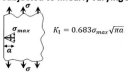

$K_I = 0.683\sigma_{max}\sqrt{\pi a}$

Penny-shaped internal crack

$K_I = \dfrac{2}{\pi}\sigma\sqrt{\pi a}$

Crack in an infinite body with crack face loading

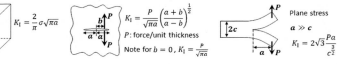

$K_I = \dfrac{P}{\sqrt{\pi a}}\left(\dfrac{a+b}{a-b}\right)^{\frac{1}{2}}$

P: force/unit thickness

Note for $b = 0$, $K_I = \dfrac{P}{\sqrt{\pi a}}$

Splitting of a rod of rectangular section

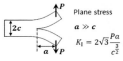

Plane stress

$a \gg c$

$K_I = 2\sqrt{3}\dfrac{Pa}{c^{\frac{3}{2}}}$

Semi-elliptical (thumbnail) crack in a semi-infinite body

$K_I = \dfrac{1.12\sigma\sqrt{\pi a}}{\Phi}$

where

a/c	Φ	a/c	Φ
0	1.000	0.6	1.277
0.1	1.016	0.7	1.345
0.2	1.051	0.8	1.418
0.3	1.097	0.9	1.493
0.4	1.151	1.0	1.571
0.5	1.211		

or $\Phi = \int_0^{\frac{\pi}{2}}[1 - \dfrac{(c^2 - a^2)}{c^2}\sin^2\theta]^{\frac{1}{2}}d\theta$

(Complete elliptical integral of the second kind)

For $a \ll c$ and $a \ll t$,

$K_I = 1.12\sigma\sqrt{\pi a}$

Middle-cracked tension (center-cracked sheet) MC(T) specimen

$K_I = \dfrac{P}{B\sqrt{W}}\sqrt{\dfrac{\pi a}{4W}}\sec\left(\dfrac{\pi a}{2W}\right)\left[1 - 0.025\left(\dfrac{a}{W}\right)^2 + 0.06\left(\dfrac{a}{W}\right)^4\right]$

Single-edge notched bend (SE(B)) specimen

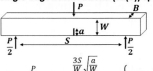

$K_I = \dfrac{P}{B\sqrt{W}}\cdot\dfrac{\dfrac{3S}{W}\sqrt{\dfrac{a}{W}}}{2(1 + 2\dfrac{a}{W})(1 - \dfrac{a}{W})^{\frac{3}{2}}}\left\{1.99 - \dfrac{a}{W}(1 - \dfrac{a}{W})\left[2.15 - 3.93\dfrac{a}{W} - 2.7\left(\dfrac{a}{W}\right)^2\right]\right\}$

Compact-tension C(T) specimen

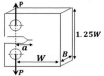

$1.25W$

$K_I = \dfrac{P}{B\sqrt{W}}\cdot\dfrac{2 + \dfrac{a}{W}}{1 - (\dfrac{a}{W})^{\frac{3}{2}}}\left[0.886 + 4.64\dfrac{a}{W} - 13.32\left(\dfrac{a}{W}\right)^2 + 14.72\left(\dfrac{a}{W}\right)^3 - 5.60\left(\dfrac{a}{W}\right)^4\right]$

FIGURE 3.3 Stress intensity K_I solutions for a number of common cracked geometries. The last three, the middle-crack tension MC(T), single-edge notched (three-point) bend SE(B), and compact-tension C(T) specimens are three of the geometries specified in ASTM Standard E1820 for K_{Ic} fracture toughness measurements.

displacements are the same (for a given mode) differing only in amplitude by the magnitude of K_i. This gives rise to the *principle of similitude* in that for two different cracks in a linear-elastic solid, it does not matter whether they are of different sizes, in different geometries subjected to different applied stresses; provided the stress-intensity factors are the same, the local stress and displacements in the vicinity of the crack tip − which invariably control fracture − will be the same as well.

LEFM represents the predictive power of a continuum mechanics approach to the problem of fracture, in that these solutions, in this case for global, isotropic, linear-elastic deformation, can equally apply to a crack in a large structure in service as well as a much smaller test sample in the laboratory, provided the K_i-fields exist in each case.[5] But the K_i-fields really must exist in each case, and this is one area where the use of fracture mechanics gets abused. In this regard, the most important issue is that the elastic stresses become infinite as $r \to 0$ at the crack tip. In reality, this will not occur as real cracks will have a finite crack-tip radius, and local in-elastic deformation will ensue at the crack tip by such processes as plastic deformation in metals, crazing in polymers, and transformation plasticity in materials that undergo an in situ phase transformation under load, e.g., in austenitic stainless steels and superelastic NiTi alloys [6]. Such local inelasticity creates what is known as the plastic zone at the crack tip [7−9].

3.1.2 Crack-tip stress triaxiality

Before we deal with the question of local plasticity, we need to consider the elastic stress state at the crack tip. When an elastic crack is subjected to a tensile load in mode I, as we have described a local tensile stress σ_{yy} is developed at the crack tip, defined by the Williams $1/\sqrt{r}$ singularity (Eq. 3.1). However, a σ_{xx} is also developed directly ahead of the crack; this stress is zero at the crack tip ($r \to 0$) where there is a free surface, but then rises with increasing r before falling to a low value far from the tip (Fig. 3.4). The origin of this stress can be envisaged by thinking of a series of infinitesimal elements ahead of the crack tip. Because of the σ_{yy} stress gradient, the element closest to the tip will be strained (in the y-direction)

---

[5] It is important to remember that fracture mechanics pertains principally to the situation of a *single dominant crack* which extends in a self-similar fashion. Certain forms of material degradation conversely involve *distributed damage,* such as the formation of cavitation at grain boundaries throughout a structure during creep deformation, the generation of fretting damage and hence small cracks, at fastener holes in an aircraft airframe or the breaking of fibers in a composite. Although these mechanisms can ultimately create a fatal crack which is amenable to fracture mechanics analyses, the structure may be severely weakened and fail due to such widespread damage. In these cases, often a more appropriate form of failure and life prediction is damage mechanics, as described in Ref. [10].

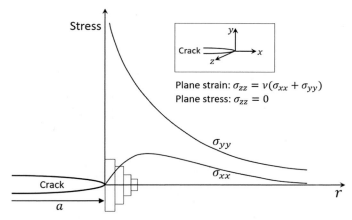

FIGURE 3.4   Elastic stresses developed at a crack tip loaded in mode I. Under plane-stress conditions, this leads to a biaxial state of stress as $\sigma_{zz} = 0$. However, under plane-strain conditions, a triaxial state of stress is created in the vicinity of the crack tip.

the most; with increasing r the next element will be strained a little less and so on as one moves away from the crack tip. However, under elastic conditions these elements will be subject to a Poisson's contraction in the x-direction, the first element at the tip contracting the most, followed by the next one, and so on. A σ_{xx} stress then must exist to maintain continuity, i.e., to prevent these element interfaces from separating so that the material behaves like a continuum.

Under plane-stress conditions where the sample thickness $B \rightarrow 0$, the through-thickness σ_{zz} stress must be zero. However, in a thick enough sample the strains cannot relax through the thickness $\varepsilon_{zz} \rightarrow 0$ such that σ_{zz} will be finite. Under such plane-strain conditions, the value of the σ_{zz} stress can be calculated using Hooke's law in terms of Young's modulus E and Poisson's ratio ν from:

$$\varepsilon_{zz} = 0 = \frac{1}{E}\left[\sigma_{zz} - \nu(\sigma_{xx} + \sigma_{yy})\right], \tag{3.3}$$

such that $\sigma_{zz} = \nu(\sigma_{xx} + \sigma_{yy})$. As ν is typically ~0.3 for most metals, and ~0.5 under plastic conditions, the magnitude of the σ_{zz} is roughly the mean of the σ_{xx} and σ_{yy} stresses. This sets up a triaxial stress state under plane-strain conditions which constitute the elastic triaxiality at the crack tip.

3.2 Crack-tip plasticity: plastic-zone size

The presence of inelasticity at the crack tip, the plastic zone (Fig. 3.5), represents a "violation" of the initial premise of LEFM, that the prevailing

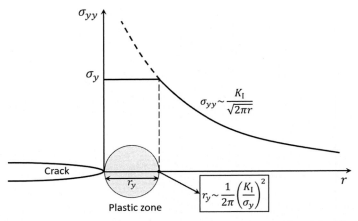

FIGURE 3.5   Idealization of the creation of a plastic-zone size, r_y, formed at the crack tip in mode I. In this simplistic derivation which is based on crack-tip stresses directly ahead of the crack tip ($\theta = 0$), the plastic zone is created at a distance $r = r_y$ from the crack tip where the local σ_{yy} crack-tip stresses exceed the yield strength σ_y.

constitutive behavior is governed by linear elasticity. We call these regions "zones of violation," as they exist for several different physical and mathematical reasons in many forms of fracture mechanics, not simply LEFM. However, we can proceed if this zone, in the present case the plastic zone, is small enough to ignore. Well, how small is small? Since the K_i-fields are planar, the *plastic-zone size*, r_y, should at least be an order of magnitude smaller than the *in-plane* dimensions of the crack size a and remaining uncracked ligament b (Fig. 3.1c). This is called a state of *small-scale yielding* (SSY) and is a mandatory requirement for the use of LEFM and the existence of crack-tip fields controlled by single parameter, K, i.e., K-dominant crack-tip conditions.

    The shape of the plastic zone can be complex though; it often has a fabiform (dogbone) shape which is dependent on whether conditions are mode I, II, or III, and whether plane stress or plane strain prevails, as discussed below. However, for the purposes of establishing whether SSY applies, we merely need an estimate of r_y. You can find many solutions for r_y in the literature [7–9], but the approximate solution for mode I cracks shown in Fig. 3.5 can readily give a value which works pretty well for most situations. We simply let the plastic zone form where the K_I-field dominated crack-tip stresses σ_{yy} directly ahead of the crack tip exceed the yield strength σ_y. As the crack-tip stresses are given by $\sigma_{yy} \sim K_I/\sqrt{(2\pi r)}$, where $\sigma_{yy} > \sigma_y$, then:

$$r = r_y \approx \frac{1}{2\pi}(K_I/\sigma_y)^2. \qquad (3.4)$$

This is not precisely correct, as it is based on a linear-elastic stress solution, and one should not simply consider conditions directly ahead of the crack tip at $\theta = 0$, but around the entire boundary of the plastic zone; also, the yield strength at the crack tip can be several times higher than the conventionally measured yield strength, σ_y, due to crack-tip constraint (this can raise the effective stress for yielding by a factor of 3 to 5 depending upon the degree of strain hardening, or even more if you believe in strain-gradient plasticity [11]). However, it is clear that the "zone of violation" for LEFM scales with $(K_I/\sigma_y)^2$, and that for the LEFM approach, based on K-dominance at the crack tip, to be "valid," a state of small-scale yielding (SSY) with $r_y << a, b$ must exist. The American Society for Testing and Materials, in its standards for fracture toughness testing, ASTM E399 [12] and E1820 [13], requires that for SSY to exist, these in-plane dimensions of a and b should exceed 2.5 $(K_I/\sigma_y)^2$, which in terms of the approximate relationship for r_y in Eq. (3.4) translates to the plastic-zone size being roughly 1/15 of the crack length and remaining uncracked ligament.

The precise shape of the plastic zone does depend on the crack displacement mode, i.e., mode I, II, or III, whether plane-stress or plane-strain conditions prevail, on the extent of strain hardening, and how the calculations were performed! There are numerous subtle differences in size and shape published in the literature [7–9,14]. One set of estimates for the actual size and shape in mode I under plane stress vs. plane strain, based on the von Mises and Tresca yield criteria, is shown in Fig. 3.6 [14]. The facts worth remembering though are that the zones, as noted above, have a fabiform shape, they are significantly larger in plane stress for a given K value, and that this is particularly noticeable directly ahead of the

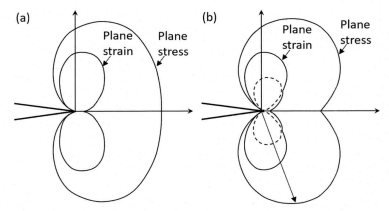

FIGURE 3.6 Estimates of the shape of the plastic zone in mode I for plane-stress and plane-strain conditions, based on the yielding criteria from (a) von Mises and (b) Tresca [14] (reproduced with permission).

crack (at $\theta = 0$). Indeed, for non-hardening plasticity conditions, the forward extent of the plastic zone in plane strain $\to 0$. They have a different shape in mode II, where they tend to have elongated peanut shape, and in mode III, where they are spherical, in both cases with the crack tip at the center of the zone. However, for most situations, you do not really need to know the exact shape of the plastic zone; the important aspect from a fracture mechanics perspective is that this zone of local plasticity, for the most commonly encountered mode I conditions, scales with $(K_I/\sigma_y)^2$ and is approximately $\dfrac{1}{2\pi}\left(\dfrac{K_I}{\sigma_y}\right)^2$ in size.

3.3 LEFM fracture criterion

3.3.1 $K_I = K_c$

Naturally the objective of defining crack-tip fields and characterizing parameters, such as the stress intensity, is to ultimately derive a usable criterion for fracture. The issue here though is that the fracture events will be motivated by the *local* stresses and strains near the crack tip, in some unspecified way, and this will change depending upon the mechanism of fracture, e.g., cleavage vs. microvoid coalescence. However, this is where the power of continuum mechanics again prevails and can lead to a *global* criterion for fracture that is essentially independent of mechanistic differences. The argument goes like this. The actual fracture event is controlled by the local stresses and strains, but for a given prevailing mode of deformation, in this case linear elasticity, these local stresses and strains are themselves characterized by the stress intensity K which is a single global parameter (involving factors that can be readily measured, such as the applied stress, crack size, and sample geometry). *Provided this K-field that defines the local stresses and strains prevails over the dimensions ahead of the crack tip where the critical fracture events occur, then we can correlate a critical value of the stress intensity, K_c, with the onset of fracture, i.e., fracture occurs when $K = K_c$,* where the value of K_c can be nominally equated to the fracture toughness of the material. As a criterion, this may seem somewhat "insipid," but it is the most honest expression of the fracture mechanics-based description of fracture. However, the "devil is in the details"!

For this simple $K = K_c$ criterion to be meaningful, the K-field must exist and properly characterize the distribution of stresses and strains at the crack tip. As noted above, this demands that the specimen size requirement of small-scale yielding (SSY) is met that the plastic zone is small enough to ignore, $r_y \ll a, b$. Although the K-field is clearly incorrect within the plastic zone, the stress and displacement fields are totally controlled by K at the plastic-zone boundary and outside the zone. If the SSY criterion is not met, then the use of LEFM and K as the characterizing parameter is simply incorrect, as it will not characterize the crack-tip stress and displacement fields, and one must resort to nonlinear-elastic fracture mechanics where the crack-tip fields are determined in the presence of some degree of inelasticity (as described in Chapter 4).

3.3.2 Plane-strain fracture

But there is another problem and that is the fact that the K-field is a planar one; it does not define the stresses and displacements in the through-thickness z-direction (Fig. 3.1), which clearly will affect fracture. The σ_{zz} stress is largely governed by the specimen thickness B, specifically in relation to the size of the plastic zone, with extremes for plane stress (where $\sigma_{zz} = 0$) and plane strain (where $\varepsilon_{zz} = 0$). A description of the deformation conditions of plane strain vs. plane stress with reference to fracture is included in Fig. 3.7 in Box 3.3. To experimentally examine how this affects the toughness, early studies (e.g., ref. [15]) on the use of LEFM looked at how the critical stress-intensity for fracture varied with this out-of-plane thickness dimension B. This is shown schematically in Fig. 3.8, where it can be seen that there is a critical thickness, B_{crit}, which is large compared to the plastic-zone size, where the fracture toughness reaches a minimum value and thereafter remains independent of B. This is where the condition of plane strain is reached such that a full triaxiality of crack-tip stresses is maintained over the majority of the crack. The measured toughness can be taken as a lower-bound and is referred to as the *plane-strain fracture toughness* K_{Ic}. In terms of toughness, K_{Ic} is the closest to being a material property in that it is independent of crack size, geometry, and thickness (as long as $B \geq B_{crit}$); it can also be considered as a realistic lower-bound for the K-based toughness. But for this value to be realized, both the conditions of SSY ($r_y \ll a, b$) and that of plane strain along the majority of the crack front ($r_y \ll B$) must be satisfied. In the ASTM E399 Standard for LEFM fracture toughness testing [12], this is articulated essentially as a single criterion:

$$B, a, b > 2.5(K_{Ic}/\sigma_y)^2, \tag{3.5}$$

for the measurement of a "valid" K_{Ic}. The precise procedures and additional criteria for the measurement of K_{Ic} in this Standard are summarized in Section 3.3.3.

BOX 3.3

Plane-strain versus plane-stress conditions

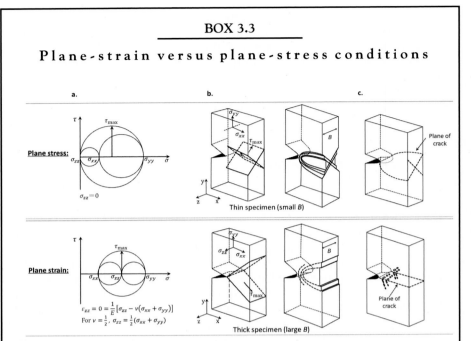

FIGURE 3.7 Definition of the deformation states of plane stress and plane strain, showing a Mohr's Circle representation, the crack-tip stresses and planes of maximum shear, and the resulting damage and crack paths. Note that the plane-strain condition confers the full stress triaxiality in the vicinity of the crack tip [14]. (a) Stress condition (close to the crack tip, σ_{yy} is in the longitudinal direction); (b) plane of τ_{max} and resulting shear deformation; (c) damage formation on planes of most extensive slip. *Adapted from D. Broek, Elementary Engineering Fracture Mechanics, fourth ed., Kluwer, New York, NY, 1991, reproduced with permission.*

3.3.3 Measurement of the plane-strain fracture toughness K_{Ic}

Procedures for conducting plane-strain fracture toughness tests to measure K_{Ic} are described in detail in ASTM Standard E399 [12]. If you read this document, it may seem like *an exercise in paranoia* as there are so many requirements, but this is because it is written to apply to a wide range of materials. Only brief outline of the salient points is given here.

One first needs to choose an appropriate test specimen. The Standard gives you the option of a middle-cracked tension (centered-cracked sheet) MC(T), single-edge notched (three-point) bend SE(B), or compact-tension C(T) specimens, although there are several variants of the latter applicable to rods and tubes. The relevant K-solutions in terms of the critical dimensions of crack size a, specimen width W, and thickness B are listed in Fig. 3.3 in Box 3.2. Ideally, to satisfy both the SSY and plane-strain criteria (Eq. 3.5), B, a, and b should all be roughly the same size;

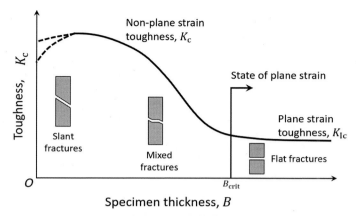

Specimen thickness, B

FIGURE 3.8 Schematic illustration of the variation in the fracture toughness, as a critical value of the stress intensity at crack initiation/instability, K_c, as a function of the out-of-plane thickness dimension B. For $B > B_{crit}$ (where $B \gg r_y$), plane-strain conditions prevail. Provided small-scale yielding conditions ($a, b \gg r_y$) additionally apply, the value of K_c above this thickness pertains to the plane-strain fracture toughness, K_{Ic}, which should be independent of crack size, geometry, and thickness (provided $B > B_{crit}$). Such plane-strain fractures are nominally flat. At smaller thicknesses, there is an increasing degree of plane-stress slant fracture (shear lips) and the resulting non-plane strain K_c toughness values tend to progressively increase as the sample thickness is reduced, except for very thin sheet ($B \to 0$) where there is presumed to be a decrease in toughness.

accordingly, $B \sim W/2$, and the starter crack should be of a length of $0.45 \leq a/W \leq 0.55$. The latter needs to be a sharp fatigue pre-crack, with requirements concerning the stress intensities that can be used to grow this crack so as not to create too large a plastic zone and hence an excessively blunted crack tip.[6]

The sample is then loaded, generally in displacement control at a desired displacement rate, as a function of typically the mouth opening displacement, which can be measured by an LVDT (linear variable differential transformer) or clip gauge mounted across the crack. The actual displacement is not used to calculate the stress intensity but nevertheless needs to be linear. As the intent is to determine the conditions for crack initiation, which ideally coincides with the fracture instability, three types of load (P)-displacement (δ), shown in Fig. 3.9, can be analyzed. In the first of these, the load rises essentially linearly until the specimen breaks; for this case, crack initiation definitely coincides with instability. The corresponding failure load P_Q, which equals the maximum load P_{max}, is then

---

[6] Specifically, the fatigue pre-cracking must be performed at a load ratio of $R = P_{min}/P_{max} \sim 0.1$ with the maximum stress intensity in the fatigue cycle initially at $K_{max} \leq 70\%\ K_Q$, where K_Q is the provisional fracture toughness, but this must be reduced to $\sim 50\%\ K_Q$ for the last half of the pre-crack length [12].

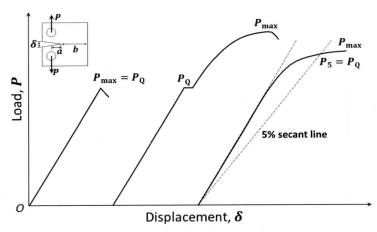

FIGURE 3.9 Types of applied load, P, vs. displacement, δ, behavior that can be analyzed in a fracture toughness test to measure K_{Ic}.

used, together with the initial pre-crack length a, to calculate a provisional toughness K_Q using the appropriate K-solution for the adopted test geometry (e.g., Fig. 3.3 in Box 3.2). In the second case, there is a "pop-in" in the load at P_Q, indicative of a burst of plane-strain (flat) fracture, before the load starts to rise again in a nonlinear fashion — the nonlinearity here is presumed to be associated with stable cracking but could also be caused by plastic deformation — prior to the sample fracturing unstably at P_{max}. The value of P_Q is again used to calculate a provisional toughness K_Q. The third case, which is often the most common, is where there is no definitive indication of crack initiation before instability occurs at P_{max}. To estimate P_Q, ASTM has devised an operational definition (akin to the definition of the offset yield strength at 0.2% plastic strain) where apparent crack initiation is defined at the point where crack extension has occurred over 2% of the remaining uncracked ligament ($\Delta a = 0.02b$). ASTM estimated that such crack extension would lead to a 5% increased compliance (lower stiffness), which they specified independent of specimen geometry. Accordingly, the construction requires you to draw a line parallel to the initial elastic line but with a 5% lower slope (5% *secant construction*); where this line intersects the load-displacement curve defines the load P_5 which is then identified as P_Q to estimate the value of K_Q.

Once P_Q is determined by one of these methods and the provisional toughness K_Q calculated, the question is whether this K_Q value is the actual plane-strain fracture toughness, K_{Ic}. This is where the specimen size requirements (validity criteria) come into play, which are often ignored yet are an essential feature of all forms of fracture mechanics. For K_I to define the stresses and displacements at the crack tip, a state of SSY must exist, i.e., $r_y \ll a, b$; additionally, for plane strain, $r_y \ll B$. As noted above, ASTM writes this as a single criterion, that $B, a, b > 2.5 \, (K_Q/\sigma_y)^2$, and additionally requires that $P_{max}/P_Q \leq 1.1$ presumably to limit the extent of plasticity and/or slow crack growth. If these criteria are met, the value of K_Q can the identified with what ASTM refers to as a "valid" plane-strain fracture toughness K_{Ic}.

When the sample thickness is limited, there is still a way to maintain the triaxiality of crack-tip stresses, i.e., plane-strain constraint, over most of the crack front and that is by using side-grooves (Fig. 3.10), which act to restrict the crack from tunneling in the center of the specimen by limiting the formation of shear lips at the plane-stress regions at the surface; this tends to lead to a relatively straight crack front. Typically, the net thickness B_N after side-grooving should be no less than ~80%B; the same K-solutions (e.g., Fig. 3.3 in Box 3.2) can still be used, but with thickness B replaced by $(B_N \cdot B)^{1\!/\!2}$.

As you can see, to select the size of your test specimen and even the fatigue pre-cracking loads, you have to know the answer before you start, i.e., you have to estimate the expected fracture toughness value to know what size of specimens to use. This unfortunately is unavoidable, but the LEFM size requirements can be very restrictive. Typical K_{Ic} values for several materials, including ceramics and polymers, are shown in Table 3.1 together with the size of test specimens needed for a *valid* result. The test methods described above actually pertain to metallic materials, although the

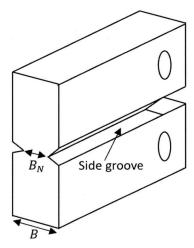

FIGURE 3.10 A compact-tension fracture specimen with side-grooves.

TABLE 3.1 Summary of the typical fracture toughness values of structural materials, their yield strength, critical crack sizes, and the size requirements to make K_{Ic} a relevant material property.

Material	Young's modulus E (GPa)	Yield strength σ_y (MPa)	Plane-strain fracture toughness K_{Ic} (MPa$\sqrt{m}$)	Plastic-zone size, r_y, at $K_I = K_{Ic}$ (mm)	Specimen size to satisfy SSY & plane strain (mm)	Critical crack size, a_{cr} at $\sigma = \sigma_y/2$ (mm)
NiCrMo 300M (Si-mod 4340) steel (200°C temper)	210	1700	60	0.2	3	1.5
NiCrMo 300M steel (650°C temper)	210	1070	~150	3	50	25
SA533B mild steel	210	500	245	40	600	300
2024-T3 Al-Cu alloy	70	370	35	1400	22	10
7075-T6 Al-Zn-Mg alloy	70	515	28	0.5	7	4
Ti-6Al-4V alloy	140	850	120	3	50	25
Silicon nitride ceramic, Si_3N_4	400	800	3–6	5×10^{-3}	0.1	0.04
Tungsten carbide, WC	410	900	10	20×10^{-3}	0.3	0.2
Diamond	1000	2800	3–7	6×10^{-4}	8×10^{-3}	4×10^{-3}
Graphene	500	130,000	3–4	1.2×10^{-7}	2×10^{-6}	1×10^{-6}
Polycarbonate	2	70	3	0.3	5	2
Human bone	15–25	~100	3–8	0.5	8	4

Unit conversion (English to SI): stress and modulus: 1 ksi $\equiv$ 6.895 MPa; stress intensity: 1 ksi$\sqrt{in}$ $\equiv$ 1.099 MPa$\sqrt{m}$.

procedures for ceramics and polymers are not too different. Polymers though do present several unique problems in toughness measurements. The interested reader may wish to review ref. [16] for some of the details.

3.3.4 Use of K as a fracture criterion in structures

The use of K_{Ic} plays an important role in providing a material parameter to characterize the fracture toughness of various materials, as listed in Table 3.1, but its other essential role is for predicting the failure of real structures. However, there are significant issues to be considered for this to work. For example, there is always the problem of the "size effect" as to whether a "coupon" test performed in the laboratory with a test sample typically the size of your finger or your hand is truly reflective of the conditions in an actual structure the size of an aircraft wing spar or a thick-section pressure vessel. One issue is whether the microstructure of the small test specimen is the same as that in the larger structure, but let us assume that we have that right. However, statistically there will always be a higher probability of imperfections and defects, e.g., pre-existing cracks, in the larger structure, and if you are trying to predict when it might fail using smooth-sample coupon tests, these much smaller lab-based test samples almost certainly would not contain the same distribution or size of defects (Leonardo da Vinci showed this in the 15th century when he found that long wires could carry less load than short ones with the same diameter). However, this is where fracture mechanics methodologies work best as they require the use of a sharp pre-cracked test coupon, which would represent a "worst-case condition" in the sample and structure; moreover, *provided all the size requirements are satisfied* (e.g., Eq. 3.5), with the use of K, the principle of similitude would imply that the critical stress intensity at fracture would apply to any sized crack in the structure.

However, as the test sample and the structure will be subjected to completely different applied (far-field) stresses, the same SSY $K = K_c$ failure criterion at a crack tip in both the sample and structure is only realistic if the K-fields, representing the local stress and displacement distributions at the crack tip over a radius of $r = r_K$, are valid, and can be superimposed on the far-field (global) stress fields; moreover, the actual distribution of stresses and displacements near any crack must truly be represented by K over dimensions comparable with the scale of fracture events. The scenario is depicted in Fig. 3.11. As the prevailing mechanism of deformation is linear elasticity, the crack-tip K-field is unique; it will apply in the test sample and structure provided it exists over dimensions large compared to the plastic-zone size, r_y (where linear elasticity clearly does not prevail), but small compared to the specimen dimensions of crack size, a, and remaining uncracked ligament,

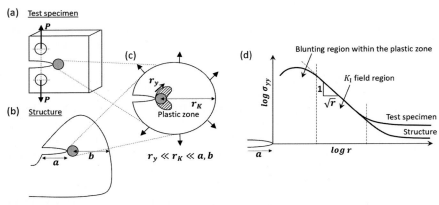

FIGURE 3.11   Application of the SSY $K = K_c$ criterion, (a) measured on a test (coupon) specimen, to (b) real structures, requires the existence of (c) a valid local K-field defining the stresses and displacements at the crack tip in both the specimen and structure. This can be achieved where the extent of the K-field, r_K, is large compared to the crack-tip plastic-zone size, r_y, yet still small compared the in-plane dimensions of crack size a and remaining uncracked ligament b (if additionally $r_y << B$, the out-of-plane component thickness, the critical stress intensity at fracture, $K_c = K_{Ic}$, will be also independent of geometry.) Despite the existence of different far-field stress fields in the test specimen and structure, (d) the local crack-tip K-field can be superimposed on these global stresses, and provided they prevail over the dimensions of fracture events, the $K = K_c$ or K_{Ic} criterion will similarly apply to the test specimen and real structure [17]. *Adapted from T.L. Anderson, Fracture Mechanics: Fundamentals and Applications, fourth ed., CRC Press, Boca Raton, FL, 2017, Reproduced by permission of Taylor and Francis Group, LLC, a division of Informa plc.*

b (for K-field dominance), i.e., $r_y << r_K << a,\ b$. If additionally, $r_y << r_K << B$, plane-strain conditions will apply, and the SSY $K = K_{Ic}$ criterion can be used, with the value of K_{Ic} being independent of geometry.

3.3.5   Flat *vs.* slant fracture surfaces

Fractures in plane strain are nominally flat. One can appreciate this from the Mohr's circle diagrams of the crack-tip stresses in Fig. 3.7 in Box 3.3. For mode I loading, the maximum local (principal) stress at the crack tip will always be σ_{yy}. Under plane-strain constraint, where the through-thickness strain is zero, as described in Eq. (3.3), the through-thickness stress σ_{zz} has a value somewhere between σ_{xx} and σ_{yy}. As σ_{yy} is the largest principal stress and σ_{xx} the smallest, the maximum shear stress at the crack tip will be located at 45° (90° in Mohr's circle space) from the y and x directions; crack-tip shear will thus be in-plane with no strain (relaxation of stress) through the thickness (plane-strain constraint) with the resulting crack advance nominally in the plane of the crack, i.e., as a flat fracture as shown. Conversely, under plane-stress conditions, the through-thickness stress σ_{zz} is zero and hence the smallest. The maximum crack-tip shear stress will now be 45° to the y and z directions such that shear and consequent stress relaxation can occur through the thickness. This starts at the side surface of the specimen as shear lips which progressively get larger as the sample thickness is reduced to eventually become a fully slant fracture, as shown in Fig. 3.12. The relaxation of plane-strain constraint from the relaxation of stresses through the thickness leads to a progressively increasing K_c toughness as the sample thickness is reduced, except for very thin sheet as the state of pure plane stress is approached ($B \rightarrow 0$) where there is limited evidence of a decrease in toughness. For the interested reader, this latter thin sheet phenomenon is modeled in Ref. [18].

3.3.6 Non fully plane-strain fracture

As noted above, if the plane-strain requirement is not met, from Fig. 3.8 it is apparent that the critical value of the stress intensity for (unstable) fracture, defined by K_c, exceeds the plane-strain value K_{Ic} and furthermore is a function of thickness; in fact, not only is this non-plane-strain-fracture toughness (sometime erroneously termed the plane-stress toughness[7]) dependent on the thickness but it is also a function of crack size and geometry — hardly a material property! We can deal with the issue of crack size using crack-resistance curves, as described in Section 3.5, but

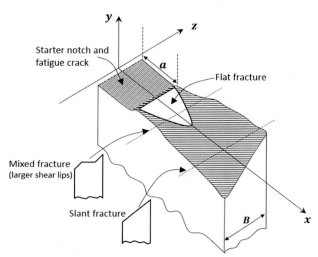

FIGURE 3.12   Creation of a slant fracture following an initial plane-strain flat fracture.

---

[7] A state of plane stress applies as $B \rightarrow 0$, and so strictly speaking the term "plane-stress toughness" only applies to the toughness of an extremely thin sheet.

physics-based models to predict how specimen thickness and geometry quantitatively affect the K_c value under non-plane-strain conditions are simply not available. Accordingly, from an engineering perspective, it is necessary to perform the K_c measurements with specimens of a thickness comparable to the intended application. Similarly, for the question of geometry, it is advisable to choose a test geometry which is most aligned with the application, although if this is not possible, it is recommended that highly constrained geometries are used, such as compact-tension or bend samples, which develop the highest triaxiality of crack-tip stresses, as these tend to give lower-bound results compared to the less constrained centered-cracked sheet MC(T) geometries.

However, there is another problem in characterizing the fracture toughness under non (or more correctly, less than fully)-plane-strain conditions. So far for SSY/plane-strain fractures characterized by K_{Ic}, which strictly represents a *crack-initiation toughness*, we have made the tacit presumption that crack initiation is synonymous with crack instability, i.e., as soon as the crack initiates from the starter notch/crack, it propagates catastrophically or at least unstably under a decreasing load (this is the basis of the engineering approximation of the 5% secant construction used in the K_{Ic} measurement standard described in Section 3.3.3). However, with the relaxation of full plane-strain constraint, the plastic-zone size, especially directly ahead of the crack, gets larger, and it is this local plasticity which stabilizes crack advance after the crack initiates, i.e., some extent of stable (slow) crack growth can ensue prior to the fracture instability. This means that it becomes difficult to characterize the fracture resistance of a material in terms of a single value toughness, because it could be defined at crack initiation, during crack growth, or at instability. Indeed, this is an issue for all fractures that are not associated with both SSY and plane strain. How one assesses the toughness under these conditions involves the use of so-called crack-resistance curves, or R-curves, which we will describe below for SSY/non-plane-strain fractures in Section 3.5 and later in Section 4.6 for non-SSY fractures in the realm of nonlinear-elastic fracture mechanics.

3.3.7 Accuracy of K_{Ic}

A brief comment on accuracy before we proceed — independent of measurement error or variability, there is an inherent uncertainty in the value of K_{Ic} for a given material/microstructure which is of the order of a little less than 10%. This largely results from the engineering approximation specified by ASTM E399 [12] to define the onset of crack initiation, i.e., the 5% secant construction to represent apparent crack extension of 2% of the remaining uncracked ligament ($\Delta a \sim 0.02b$). Recall that the local crack-tip stresses, e.g., σ_{yy}, are defined by the first term in the Williams singular $1/\sqrt{r}$ solution in Eq. (3.1a,b), i.e., as $r \to 0$, in order for K to represent a singular characterization of the crack-tip fields. To be able to do this, all higher order terms in the $1/\sqrt{r}$ solution need to be small enough to ignore. However, as one moves ahead of the crack tip for finite r, the higher order terms become more relevant such that the actual crack-tip stresses will deviate from that given by the first term of the K-solution. The problem is that this deviation from the (first-term) K-field definition of the local stresses and displacements is a function of specimen geometry. If the true value of σ_{yy}, for example, is calculated from the full expansion of Eq. (3.1a,b) and compared to the value derived from just the first term, $\sigma_{yy}(K)$, the error in the σ_{yy} stress field from using $\sigma_{yy}(K)$ will be positive for the compact-tension C(T) and the single-edge notched bend SE(B) and tension SE(T) geometries, yet negative for the middle-cracked tension MC(T), as shown in Fig. 3.13. This leads to a ~7% variation in the value of K_{Ic} measured using the 5% secant construction for the various different specimen geometries specified in the ASTM Standard. We will discuss this further in Section 4.7 when we talk about the T-stress.

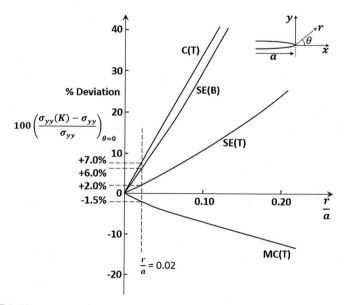

FIGURE 3.13 Variation in the crack-tip stress σ_{yy} calculated from the first (K-based) term in the series expansion in Eq. (3.1a,b), $\sigma_{yy}(K)$, compared to that calculated from the full series expansion, σ_{yy}, expressed as a percentage, $100[(\sigma_{yy}(K) - \sigma_{yy})/\sigma_{yy}]_{\theta = 0}$, as a function of distance ahead of the crack tip. Note that the deviation from the pure crack tip K-field with increasing r differs markedly in magnitude and sign for different specimen geometries [18] (reproduced with permission).

3.3.8 Relevance of K_{Ic}

Much like the yield and tensile strengths, the fracture toughness is important in the choice of a material in design. In this respect, the concept of the *fracture mechanics triangle* (Fig. 3.14) can be useful. As $K_{Ic} \sim Y\sigma^\infty \sqrt{\pi a_c}$, the scenario is that as the design of a component will essentially specify the geometry and applied stress, $Y\sigma^\infty$, and the choice of material will decide the toughness, K_{Ic} (if this is the appropriate parameter governing its fracture), this will result in a *critical crack size*, a_c, which is the largest defect or crack that this component can tolerate without fracture at that particular level of applied stress, i.e.,

$$a_c \approx \frac{1}{\pi} \left(\frac{K_{Ic}}{Y\sigma^\infty} \right)^2 . \tag{3.6}$$

Values of a_c for different materials, estimated at an applied stress of $\sigma_y/2$, are listed in Table 3.1.

The critical crack size is an essential aspect of the design of safety critical structures, where the safe lifetime is determined by how long it takes a pre-existing defect to grow, by subcritical crack growth mechanisms such as fatigue, environmentally-assisted cracking or creep, to failure, e.g., where $K = K_c$ or K_{Ic}. Since one or more of these subcritical cracking mechanisms are inevitable in most structures, the value of a_c represents the size of defect that must be detected to prevent catastrophic failure. In the aluminum fuselage of commercial aircraft, a_c can be of the order of fractions of a meter, whereas for turbine blades in military jet engines a_c can be significantly smaller, even in the millimeter range. This makes the periodic nondestructive evaluation (NDE) of components to detect incipient cracks, particularly for serious safety-critical applications

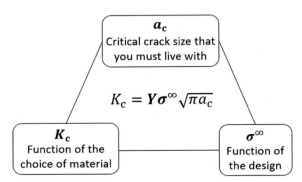

FIGURE 3.14   The "fracture mechanics triangle", relating the applied stress σ^∞, which is fixed by the design of a component, the fracture toughness K_c, which is fixed by the choice of material, and the resulting critical crack size a_c, which is the largest crack that the component can tolerate without fracturing at that stress level.

such as aerospace and nuclear structures, so vital for their safe operation. It is also the reason why monolithic ceramics are not used as gas turbine components; although they are high strength and lightweight with excellent high-temperature resistance, their critical crack sizes can be measured in tens of micrometers and are highly sensitive to stress (see also Section 7.4).

As it is independent of crack size and geometry (provided plane-strain conditions prevail), the value of K_{Ic} can serve as a reproducible material property to compare the relative resistance to fracture of different materials, as in Table 3.1. However, a more important issue is whether K_{Ic} (or equivalent K_{IIc} and K_{IIIc}) values are actually useful, or even relevant, for a material component in a given application. For example, for the ultrahigh-strength (1700 MPa), quenched and 200°C tempered, AISI 4340 steel listed in Table 3.1, K_{Ic} is applicable to all section sizes larger than a few millimeters; as one application of this martensitic steel is for aircraft landing gear, its K_{Ic} value of ~60 MPa√m is a relevant parameter, which further could be readily measured in the laboratory. In contrast, the lower-strength (500 MPa) mild steel (SA533B) has a K_{Ic} value of over 200 MPa√m, meaning that SSY/plane-strain conditions are only met in section sizes of >0.6 m; this may be relevant to its major use as a thick-section nuclear pressure vessel steel, but few laboratories in the world could ever test such huge specimens (this incidentally was the prime motivation to develop nonlinear-elastic fracture mechanics, as described in Chapter 4). Finally, solution treated and aged aluminum alloys, such as 2024 and 7075, require section sizes in the range of tens of millimeters to achieve SSY and plane-strain conditions. A major use of these alloys is as the fuselage of commercial aircraft which are typically only a few millimeters thick; consequently, K_{Ic} is simply not a pertinent parameter (except perhaps as a lower bound) for defining their resistance to fracture for this significant application.

Since the validity of a K_{Ic} value is attained by having a small plastic zone to meet nominally elastic conditions, which is only achieved in low toughness or high strength materials, this prompted the facetious, yet famous, quote by Joe Gallagher of the U.S. Air Force to the effect that *if you can ever unambiguously characterize the fracture toughness of a material by K_{Ic}, you probably don't want to be using that material!*

3.4 *G*-based energy approach

3.4.1 Definition of *G*

In addition to the *K*-based characterizing parameter approach described above, there is an alternative LEFM methodology which provides an energy-based approach for fracture. First proposed by George

Irwin [19],[8] and nominally equivalent (although far less restrictive) than the Griffith criterion (Eq. 3.7), it involves the concept of the *strain energy release rate*, G, which exceeds a critical value, $G = G_c$, at the onset of fracture. For a cracked linear-elastic body, G is defined as the rate of change in potential energy, U_{PE}, per unit increase in crack area, A, viz.,

$$G = -\frac{dU_{PE}}{dA}, \tag{3.7}$$

where U_{PE} can be defined in terms of the stored strain energy, U_ε, and the work done, W_D, by the external forces, P, by:

$$U_{PE} = U_\varepsilon - W_D. \tag{3.8}$$

To compute G, we consider a plate of thickness B containing an edge crack of length, a, subjected to either displacement or load control. For a fixed displacement Δ (Fig. 3.15a), the work done will be zero, $W_D = 0$, viz.,

$$U_{PE} = U_\varepsilon = \int_0^P \Delta dP = \frac{\Delta P}{2}, \tag{3.9}$$

such that the strain energy release rate is given by:

$$G = -\frac{1}{B}\left(\frac{dU_{PE}}{da}\right)_\Delta = -\frac{\Delta}{2B}\left(\frac{dP}{da}\right)_\Delta = -\frac{P^2}{2B}\left(\frac{dC}{da}\right)_\Delta, \tag{3.10}$$

where C is the *specimen compliance* (the inverse of the plate stiffness) given by $C = \Delta/P$. Note that with crack extension da, under fixed displacement the net strain energy decreases.

An essentially identical result is obtained under load control. If the plate is instead subjected to a fixed load P (Fig. 3.15B), the work done will be finite such that W_D and U_ε will be given by:

$$W_D = P\Delta \quad \text{and} \quad U_\varepsilon = \int_0^\Delta Pd\Delta = \frac{P\Delta}{2}. \tag{3.11}$$

---

[8] George R. Irwin is widely considered to be the "Father of Fracture Mechanics." He presented the G-based approach in the mid-1950s whereas the K-based methodology at that time was favored in the Soviet Union. Both forms of LEFM, which are perfectly complementary, and are widely used today, with the metals and ceramics fracture communities invariably utilizing a K-based approach, whereas the polymer (and mixed-mode) communities often preferring the use of G. The use of an italic typeface for G in the style of the early Italian cursives was originally proposed by Irwin and persists today.

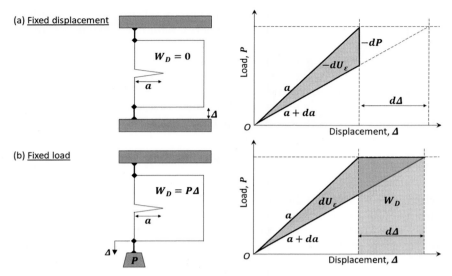

FIGURE 3.15 A linear-elastic plate containing a crack of length a "loaded" (a) at a fixed displacement $\varDelta$, and (b) at a fixed load P, relevant to the definition of the strain-energy release rate G, in Eqs. (3.8)–(3.13).

As from Eq. (3.9), $U_{PE} = -P\varDelta/2 = -U\varepsilon$, G is now given by:

$$G = \frac{1}{B}\left(\frac{dU_\varepsilon}{da}\right)_P = \frac{P}{2B}\left(\frac{d\varDelta}{da}\right)_P = +\frac{P^2}{2B}\left(\frac{dC}{da}\right)_P, \qquad (3.12)$$

such that crack extension da causes an increase in strain energy. Comparing Eqs. (3.10) and (3.12), one can see that the strain-energy release rate G is the same under either displacement or load control, differing only in sign and the incremental amount $\frac{1}{2}dPd\varDelta$.

In the early days of fracture mechanics, Eqs. (3.10) and (3.12) were used to experimentally determine G-solutions in order to calibrate various test geometries for fracture toughness measurements; the procedure involved extending a crack and monitoring the load-displacement relationship after each crack increment to determine how the compliance varied with crack size, dC/da.

3.4.2 Characterizing parameter *vs.* energy release rate approach

We thus have two seemingly distinct approaches to characterizing the growth of cracks in elastic solids: (i) the *characterizing parameter, K-based, approach*, where the stress intensity is used to define the *local* stress and displacement fields in the vicinity of a crack tip, and (ii) the *energy,*

G-based, approach, where the *global* strain-energy release rate is determined to define the net change in potential energy with crack extension. Irwin, however, through a very clever thought experiment [20], was able to demonstrate that the two parameters are intimately related. For a mode I crack of length $a + \Delta a$, he applied a compressive force to close the crack to a length a and calculated the energy release rate G from the work done. From the change in crack-opening displacements, he was also able to deduce the K_I at the original crack tip. By comparing the two scenarios, he was able to show for mode I that:

$$G = \frac{K_I^2}{E'},$$
(3.13)

where $E' = E$, Young's modulus in plane stress, and $E/(1 - v^2)$ in plane strain (v is Poisson's ratio).

By considering all three modes of crack displacements, the general relationship between G and K_I, K_{II}, and K_{III} can similarly be determined and is given, in terms of E' and the shear modulus μ, by:

$$G = \frac{K_I^2}{E'} + \frac{K_{II}^2}{E'} + \frac{K_{III}^2}{2\mu}.$$
(3.14)

These equivalence equations are important, not only for the conversion of G to K_i values, but also for multiaxial loading where G can be used as a mixed-mode, crack-driving force. This is discussed further in Section 3.6 on mixed-mode fracture.

3.5 Crack-resistance R-curves

As described in Section 3.3, fractures under SSY and plane-strain conditions tend to involve fracture instability at, or essentially immediately after, crack initiation. However, with the relaxation of plane-strain constraint with thinner geometries approaching plane-stress conditions, the increasing degree of crack-tip plasticity tends to stabilize cracking such that stable crack advance occurs prior to fracture instability. This is also the case for non-SSY (nonlinear-elastic) conditions, in plane strain or plane stress, where the extent of local plasticity is even larger. Such behavior can be described in terms of the *crack-resistance curve*, or R-curve.

So what is a R-curve? Let us first consider a simple center-cracked plate containing a middle crack of length $2a$ subjected to an applied tensile stress σ^∞ (Fig. 3.16A). For simplicity, we will assume the infinite plate approximation, with SSY but under non-plane-strain conditions:

$$K_I = \sigma^\infty \sqrt{\pi a}, \quad \text{such that } G = K_I^2/E = \frac{(\sigma^\infty)^2 \pi a}{E}.$$
(3.15)

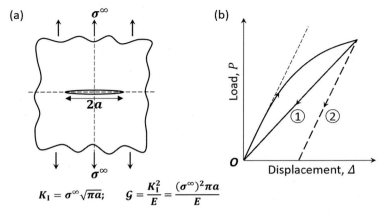

$$K_{\mathrm{I}} = \sigma^{\infty}\sqrt{\pi a}; \qquad \mathcal{G} = \frac{K_{\mathrm{I}}^2}{E} = \frac{(\sigma^{\infty})^2 \pi a}{E}$$

FIGURE 3.16 (a) A center-cracked sheet MC(T) samples loaded in tension under conditions of SSY but non-plane strain, and (b) resulting nonlinear load-displacement curves. By examining the unloading curves it is possible to discern the reason for the nonlinearity: unloading path 1 represents 100% crack growth, unloading path 2 represents 100% plastic deformation. See text for explanation.

The corresponding load/load-line displacement (P-Δ) plot is shown in Fig. 3.16b, where it is apparent that the loading line is nonlinear, i.e., it deviates from the (dashed) linear-elastic loading line at larger Δ. There are two possible reasons for this nonlinearity — stable crack extension and/or local plasticity — which can be deduced by unloading (which, of course, is always elastic). Unloading path 1 is due to crack extension: the reciprocal unloading slope, which is indicative of the specimen compliance, is larger meaning that the crack is longer, and the unloading curve returns to the origin meaning that there is no residual strain due to plasticity. Unloading path 2, conversely, is purely due to plasticity; the unloading and loading slopes are the same, meaning the compliance and hence crack size is unchanged, but there is a residual strain at zero load indicative of plasticity deformation. As we are considering the case of small-scale yielding here, we must presume then that the nonlinearity in this case is principally due to stable crack extension.

From this load-displacement curve, as we know the load and crack size (i.e., the initial crack size, a_o, plus any stable crack extension Δa) at each point, we can calculate the LEFM "crack-driving force," e.g., the applied $\mathcal{G}$ or K_{I}, in terms of the crack size, $a = a_o + \Delta a$.[9] In this example, we will use $\mathcal{G}$ (as it makes the graphical construction a little easier). The resulting $\mathcal{G}$-R

[9] Although in LEFM we use K_{I} or $\mathcal{G}$ as the crack-driving force for the crack resistance, the same procedure is also used to develop R-curves in nonlinear-elastic fracture mechanics but with the J-integral or the CTOD as the characterizing parameters. This is described in, respectively, Chapters 4 and 5.

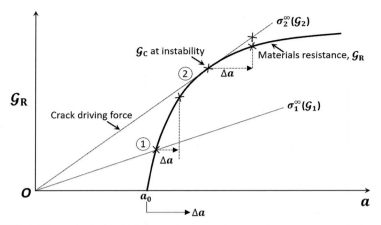

FIGURE 3.17 $\mathcal{G}_R(\Delta a)$ crack-resistance R-curve, showing the construction to determine the fracture instability at $\mathcal{G}_c$.

curve is shown in Fig. 3.17; it represents a measure of resistance to cracking, $\mathcal{G}_R$ (sometimes simply termed R) as a function of the crack extension. Whereas in SSY/plane strain, we can measure a single value parameter to define the fracture resistance — $\mathcal{G}_{Ic}$ or K_{Ic} (depending on how you calculate it) — now, under non-plane-strain conditions, it is the entire R-curve — $\mathcal{G}R(\Delta a)$ or $K_R(\Delta a)$ — which defines the material resistance to fracture.

But how does it work? For SSY/plane-strain conditions, we apply stress σ^∞ to the sample and compute the applied crack-driving force — $\mathcal{G}$ or K_I — if it exceeds the fracture toughness — $\mathcal{G}_{Ic}$ or K_{Ic} — the material fractures unstably, as described in Section 3.3. Now when we apply a $\mathcal{G}$ or K_I to the sample, from Eq. (3.15), it appears as a line from the origin; for the $\mathcal{G}$-R curve in Fig. 3.17, as a straight line $\mathcal{G} = \{(\sigma^\infty)^2 \pi a\}/E$. If we apply an initial (far-field) stress of σ_1^∞, equivalent to an applied driving force $\mathcal{G}_1$, it will intersect the R-curve at point 1, $\mathcal{G} = \mathcal{G}_R$, and so at that stress the crack will grow. If it extends by an amount Δa, and we then check the magnitudes of the applied $\mathcal{G}$ and $\mathcal{G}_R$, it is clear that $\mathcal{G} < \mathcal{G}_R$ meaning that the crack will require additional stress (and $\mathcal{G}$) to be able to advance, i.e., it can grow but stably. Thus, for stable crack growth:

$$\mathcal{G} = \mathcal{G}_R \quad \text{and} \quad d\mathcal{G}/da \leq d\mathcal{G}_R/da. \tag{3.16}$$

This will be the case for all further increases in applied stress until the line representing σ_2^∞, which intersects at point 2 *at a tangent* to the $\mathcal{G}_R$ curve; now if the crack extends by any amount Δa, $\mathcal{G} > \mathcal{G}_R$, meaning that the crack does not require an increasing driving force to extend and so will grow unstably. Thus, the non-plane-strain fracture toughness $\mathcal{G}_c$

(or equivalently K_c) can be defined at this point of fracture instability, when:

$$dG/da > dG_R/da. \tag{3.17}$$

The question that one needs to ask now is whether this SSY/non-plane-strain fracture toughness G_c (or equivalently K_c) is a material property. As mentioned in Section 3.3, the SSY/plane-strain G_{Ic} *or* K_{Ic} *values are independent of crack size, geometry, and sample thickness* (provided plane-strain constraint prevails), but the SSY/non-plane-strain G_c or K_c values are unfortunately not.

So taking each of these factors in turn, what about the effect of crack size? Shown in Fig. 3.18 are SSY R-curves in (a) plane strain and (b) non-plane strain. The first thing to notice is that the R-curve measured at one particular initial crack size, e.g., a_1, and then readily shifted to another initial crack size, here a_2. This is because R-curves are not dependent on the crack size a but on the extent of crack extension Δa. That is why they

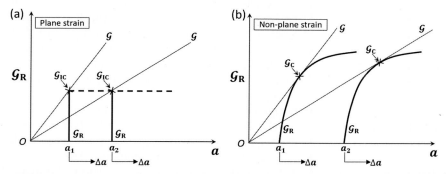

FIGURE 3.18   R-curves in small-scale yielding showing (a) a plane-strain situation where the G_c at fracture instability is independent of crack size, and (b) a non-plane-strain situation where the value of G_c is function of crack size a.

are referred to as $G_R(\Delta a)$ or $K_R(\Delta a)$. The reason for this is complex but in metallic materials results mechanistically from the fact that plasticity is the prime factor that stabilizes crack growth, particularly as the crack advances and leaves a wake of plastic zone.[10]

Given that the R-curve is a constant that can be shifted to any specific initial crack size, if we examine Fig. 3.18a, the R-curves for SSY/plane-strain conditions where crack instability occurs at crack initiation without significant stable cracking, the point of tangency at $G = G_{Ic}$ (or $K_I = K_{Ic}$) is always the same, and independent of crack size a. In contrast, for SSY/non-plane-strain conditions in Fig. 3.18b, where rising R-curves are apparent due to the occurrence of stable crack growth, although the R-curve is the same, the point of tangency will change with the initial crack size, such that the value of G_c (or K_c) will be different. Thus, the SSY/non-plane-strain G_c and K_c fracture toughness will not be independent of crack size a (like the plane-strain G_{Ic} and K_{Ic} values are), but since the R-curves are independent of a, this can be taken into account.

But what about geometry and sample thickness? We have described above how these G_c or K_c values vary with thickness B, as shown in Fig. 3.8, and geometry. The only solution is to measure the R-curve in a sample pertaining to the actual thickness of the service component that you are trying to analyze and, if the loading geometry of the pertinent flaws is not known, to use a highly constrained test geometry, such as the deep-cracked bend SE(B) or compact-tension C(T) samples, to give a lower-bound for the R-curve.

There is a final point though concerning the onset of fracture instability. The tangent construction described above applies specifically to a load-controlled test. Under displacement control, depending on the crack size and geometry, the load can drop with crack extension. As the crack-driving force may therefore be reduced with crack extension, the instability point is therefore delayed in displacement-controlled tests and crack growth becomes more stable.

---

[10] To explain this in simple terms, we have seen that the strain distribution ahead of a crack tip under linear-elastic deformation conditions scales as $1/\sqrt{r}$, where r is the distance ahead of the crack tip [3] (Eq. 3.1). Inside the plastic zone though, nonlinear-elastic solutions give the local strain distribution ahead of the crack-tip scaling approximately with $1/r$ [21,22]. Both these solutions pertain to stationary cracks with no plastic wake. If the corresponding elastic-plastic crack-tip fields are examined for nonstationary cracks which actually move and leave a trail of plastic zone in their wake, the local strain distribution is even weaker and scales with $\ln(1/r)$ [23]. This means that as a crack advances at a constant applied driving force (G or K_i), the local strain in the vicinity of the crack tip is reduced due to this wake plasticity; consequently, to sustain crack growth the applied driving force has to be increased which results in the rising R-curves depicted in Figs. 3.17 and 3.18b.

Summarizing for small-scale yielding conditions, whereas the plane-strain fracture toughness, either G_{Ic} or K_{Ic}, is close to being a material property, this cannot be said for the corresponding non-plane-strain G_c or K_c values which need to be measured in the relevant sample thickness and geometry, but can be corrected for the pertinent crack size by using the R-curve construction described above.

3.6 Mixed-mode fracture

Mixed-mode fracture mechanics involves situations where a cracked specimen or structure is subjected to applied (far-field) multiaxial loads. It is also relevant under pure tensile (mode I) loading where a crack is located at an inclined angle to the principal stress axes, where a small section of a crack is deflected or twisted, for example from a mode I path, or where a crack is located in an interface between two dissimilar materials. In all these situations, the crack will be subjected to combinations of K_I, K_{II}, and/or K_{III}. Before discussing how one can deal with combinations of different modes to determine a viable mixed-mode crack-driving force, we will briefly examine some of these cases where a seemingly mode I crack can experience mixed-mode loads.

3.6.1 Inclined cracks

Firstly, an inclined center crack in an infinite (thin) plate under a pure tensile stress σ_{yy}^{∞} (Fig. 3.19a) will be subjected to both a K_I and a K_{II}. If the

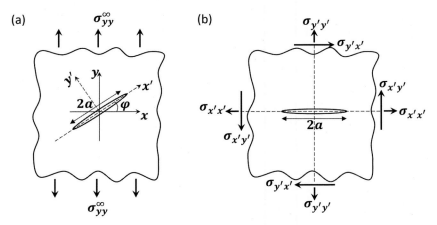

FIGURE 3.19   (a) Inclined center-crack in an infinite sheet under far-field tension; (b) transfer of the x-y axis to the x'-y' axes in order to determine the K_I and K_{II} stress intensities.

crack, of length $2a$, is inclined at a finite angle φ to the normal to the stress axis, the K_I and K_{II} values can be readily computed by transforming the original x, y stress axes to x', y' axes that are parallel and normal to the crack plane. Using the tensor transformation rule, or Mohr's circle, to calculate the new $\sigma_{y'y'}$ and $\sigma_{x'y'}$ stresses, the tensile $\sigma_{y'y'}$ stress will generate a K_I and the shear $\sigma_{x'y'}$ stress a K_{II}, given by:

$$K_I = \sigma_{y'y'} \sqrt{\pi a} = \sigma_{yy}^{\infty} \cos^2\varphi \sqrt{\pi a},$$
$$K_{II} = \sigma_{x'y'} \sqrt{\pi a} = \sigma_{yy}^{\infty} \sin \varphi \cos \varphi \sqrt{\pi a}. \tag{3.18}$$

3.6.2 Deflected cracks

A second example where a crack growing under a pure mode I tensile stress can undergo mixed-mode loading is when a small segment at its tip is deflected off the plane of maximum tensile stress; an in-plane tilt will generate a mode II K whereas an out-of-plane twist will additionally generate a mode III K. This incidentally is a powerful (extrinsic) mechanism of toughening when cracks are deflected, e.g., at hard particles/ phases or crystallographically at grain boundaries, as the resulting deviation in crack path acts to lower the crack-driving force (see also Section 6.4). Let us consider a crack of length a, subjected to an applied far-field tensile (and shear stress, for completeness), i.e., subjected to a global K_I and K_{II}, which undergoes a small, in-plane, deflection of length a', where $a' << a$, through an angle φ to the crack plane (Fig. 3.20). The deflection will result in both a local mode I and mode II stress intensity being developed at the crack tip (even if the global $K_{II} = 0$). Redefining the x-y axes at the tip of the kink, we can calculate the global σ_{yy} and σ_{xy} to determine these local stress intensities; they are generally, respectively, written as k_1 and k_2, and can be expressed as [24]:

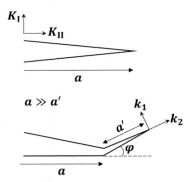

FIGURE 3.20   Determination of the local stress intensities k_1 and k_2 for a deflected crack (in-plane tilt through angle, φ) subjected to global stress intensities of K_I and K_{II}.

$$k_1 = a_{11}(\varphi)\, K_{\mathrm{I}} + a_{12}(\varphi)\, K_{\mathrm{II}},$$

$$k_2 = a_{21}(\varphi)\, K_{\mathrm{I}} + a_{22}(\varphi)\, K_{\mathrm{II}}, \tag{3.19}$$

where the factors a_{ij} are a function of the nature of the kink and the deflection angle φ. For the in-plane tilt shown in Fig. 3.20 [25,26]:

$$a_{11}(\varphi) = \cos^3\left(\varphi/2\right),$$

$$a_{12}(\varphi) = -3\sin\left(\varphi/2\right)\cos^2\left(\varphi/2\right),$$

$$a_{21}(\varphi) = \sin\left(\varphi/2\right)\cos^2\left(\varphi/2\right),$$

$$a_{22}(\varphi) = \cos\left(\varphi/2\right)\left[1 - 3\sin^2\left(\varphi/2\right)\right]. \tag{3.20}$$

3.6.3 Interface cracks

A far more complicated example of where mixed-mode loading can be induced under seemingly pure mode I conditions is a crack that sits at the interface between two dissimilar elastic materials in a tensile loaded specimen (Fig. 3.21). Due to the difference in elastic properties, e.g., the

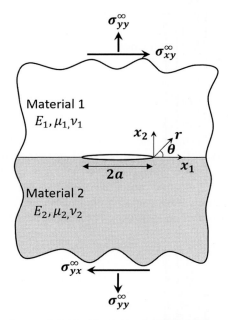

FIGURE 3.21 Interface crack located between two dissimilar materials: material 1 with elastic properties E_1, μ_1, ν_1, and material 2 with elastic properties E_2, μ_2, ν_2, subjected to far-field loading.

Young's modulus E, shear modulus μ, and Poisson's ratio ν, across the interface, an interfacial crack is subjected to shear as well as tensile stresses, i.e., K_{II} as well as K_I, even under pure mode I loading. Specifically, the ratio of normal to shear stresses ahead of a crack differs markedly from that expected from the far-field loading, and can be quantified in terms of the Dundurs' parameters [27], one of which is $\beta = \frac{1}{2}\{[\mu_1(1 - 2\nu_2) - \mu_2(1-2\nu_1)]/[\mu_1(1 - 2\nu_2) + \mu_2(1 - 2\nu_1)]\}$, where the subscripts refer to material 1 and 2. The linear-elastic stress field, at distance r ahead of the crack,[11] is characterized by a complex stress intensity factor $K = K_I + iK_{II}$, where the real part is the mode I term and the imaginary part is the mode II term ($i = \sqrt{-1}$), and is given by Refs. [28,29]:

$$\sigma_{ij} = \frac{Re\left[K\, r^{i\in}\right]}{\sqrt{2\pi r}} \overset{I}{\underset{ij}{\sum}}(\theta, \in) + \frac{Im\left[K\, r^{i\in}\right]}{\sqrt{2\pi r}} \overset{II}{\underset{ij}{\sum}}(\theta, \in), \qquad (3.21)$$

where $\in$ is the so-called bimaterial constant:

$$\in \; = \; \frac{1}{2\pi}ln\left\{\frac{1 - \beta}{1 + \beta}\right\}, \qquad (3.22)$$

and $\overset{I}{\underset{ij}{\sum}}(\theta, \in)$ and $\overset{II}{\underset{ij}{\sum}}(\theta, \in)$ are simply the mode I and mode II angular functions, which both equal unity along the interface ahead of the crack tip ($\theta = 0$). Accordingly, this complicated solution for stresses ahead of an interface crack presents another example of a crack subjected to solely far-field tensile loads, which nevertheless generates mixed-mode stress intensity factors.

3.6.4 Mixed-mode crack-driving force

In situations where a crack is subjected to combined K_I, K_{II}, and/or K_{III}, whether due to tension plus shear/twisting far-field loads or, as described above, from inclined, deflected, or interfacial cracking, mixed-mode stress intensities cannot be superposed, as the principle of superposition can only be used for Ks of the same mode. However, as energies can be readily added, Eq. (3.14), which is reproduced again here as Eq. (3.23), presents $\mathcal{G}$ as a viable mixed-mode crack-driving force for the self-similar extension

---

[11] For this complex bimaterial interface crack solution, the singular stresses become oscillatory and the crack surfaces become interpenetrating as $r \to 0$ at the crack tip. As this is not physically realistic, this region (which is invariably very small) represents another "zone of violation" similar to that of the crack-tip plastic zone in the linear-elastic singularity solution for homogeneous materials.

of a crack-subjected tensile, shear, and anti-plane shear displacements under linear-elastic conditions:

$$\mathcal{G} = \frac{K_I^2}{E'} + \frac{K_{II}^2}{E'} + \frac{K_{III}^2}{2\mu}. \tag{3.23}$$

3.6.5 Crack paths

Before we leave the question of mixed-mode loading, another relevant issue is the question of the crack path. Crack trajectories are invariably governed by two distinct criteria: that of the path of maximum mechanical crack-driving force and that of the weakest microstructural resistance. If these two criteria are incommensurate in a material under load, this generally leads to a high toughness as the crack is driven in different directions, whereas if they are commensurate the toughness is generally much lower [30].

Of course, mechanics can only address the issue of a maximum driving force, which under mode I linear-elastic conditions will be the path of maximum tangential stress, i.e., maximum tensile stress perpendicular to the crack path; this is the well-known Erdogan and Sih criterion [31]. In terms of LEFM, this is often expressed as the crack following a $K_{II} = 0$ path, i.e., dominated by the first term in the series expression for the local stress (e.g., Eq. 3.1a). Indeed, lacking major microstructural effects, this is generally the case for most brittle fractures. However, a more general criterion for multi-axial loading is one of a path of maximum $\mathcal{G}$ [32], which under linear-elastic conditions is almost identical to the $K_{II} = 0$ path, except for a small difference at high degrees of shear loading.

References

[1] H.M. Westergaard, Bearing pressures and cracks, J. Appl. Mech. 6 (1939) 49.
[2] I.N. Sneddon, The distribution of stress in the neighbourhood of a crack in an elastic solid, Proc. Roy. Soc. (London). A 187 (1946) 229.
[3] M.L. Williams, On the stress distribution at the base of a stationary crack, J. Appl. Mech. 24 (1957) 109.
[4] H. Tada, P.C. Paris, G.R. Irwin, The Stress Analysis of Cracks Handbook, third ed., ASME, New York, NY, 2000.
[5] Y. Murakami ed, Stress Intensity Factors Handbook, vols. 1–3, Soc. Matls. Sci. Japan & Pergamon, Kyoto, Japan. (1987, 1992).
[6] A. Das, S. Tarafder, Experimental investigation on martensitic transformation and fracture morphologies in austenitic stainless steel, Int. J. Plast. 25 (2009) 2222.
[7] G.R. Irwin, Plastic zone near a crack and fracture toughness, Proc. 7th Sagamore Conf. 4 (1961) 63.
[8] D.S. Dugdale, Yielding in steel sheets containing slits, J. Mech. Phys. Solid. 8 (1960) 100.

[9]   G.I. Barenblatt, The mathematical theory of equilibrium cracks in brittle fracture, Adv. Appl. Mech. VII (1962) 55.

[10] D. Krajcinovic, Damage mechanics, Mech. Mater. 8 (1989) 117.

[11] J.Y. Chen, Y. Wei, Y. Huang, J.W. Hutchinson, K.C. Hwang, The crack tip fields in strain gradient plasticity: the asymptotic and numerical analyses, Eng. Fract. Mech. 64 (1999) 625.

[12] ASTM Standard E399-20, Standard Test Method for Linear-Elastic Plane-Strain Fracture Toughness of Metallic Materials, in: Annual Book of ASTM Standards, vol. 3.01, American Society for Testing and Materials, West Conshohocken, PA, 2020.

[13] ASTM Standard E1820-20, Standard Test Method for Measurement of Fracture Toughness, in: Annual Book of ASTM Standards, vol. 3.01, American Society for Testing and Materials, West Conshohocken, PA, 2020.

[14] D. Broek, Elementary Engineering Fracture Mechanics, fourth ed., Kluwer, New York, NY, 1991.

[15] J.G. Kaufman, F.G. Nelson, More on specimen size effects in fracture toughness testing, in: ASTM STP 559, Am. Soc. Test. Mater. (1973) 74–98. West Conshohocken, PA.

[16] L.G. Malito, J.V. Sov, B. Gludovatz, R.O. Ritchie, L.A. Pruitt, Fracture toughness of ultra-high molecular weight polyethylene: a basis for defining the crack-initiation toughness in polymers, J. Mech. Phys. Solid. 122 (2019) 435.

[17] T.L. Anderson, Fracture Mechanics: Fundamentals and Applications, fourth ed., CRC Press, Boca Raton, FL, 2017.

[18] J.F. Knott, Fundamentals of Fracture Mechanics, Butterworths, London, UK, 1973.

[19] G.R. Irwin, Onset of fast crack propagation in high strength steel and aluminum alloys, Proc. Sagamore Res. Conf. 2 (1956) 289.

[20] G.R. Irwin, Analysis of stresses and strains near the end of a crack traversing a plate, J. Appl. Mech. 24 (1957) 361.

[21] J.R. Rice, G.F. Rosengren, Plane strain deformation near a crack tip in a power-law hardening material, J. Mech. Phys. Solid. 16 (1968) 1.

[22] J.W. Hutchinson, Singular behavior at the end of a tensile crack tip in a hardening material, J. Mech. Phys. Solid. 16 (1968) 13.

[23] J.R. Rice, W.J. Drugan, T.-L. Sham, Elastic-plastic analysis of growing cracks, in: ASTM STP 700, Am. Soc. Test. Mater. (1980) 189.

[24] B.A. Bilby, G.E. Cardew, I.C. Howard, Stress intensity factors at the tips of kinked and forked cracks, in: Analysis and Mechanics. Proc. 4th Intl. Conf. on Fracture, Pergamon, Oxford, UK, 1978, p. 197.

[25] B. Cotterell, J.R. Rice, Slightly curved or kinked cracks, Int. J. Fract. 16 (1980) 155.

[26] K.T. Faber, A.G. Evans, Crack deflection processes − I. Theory, Acta Metall. 31 (1983) 565.

[27] J.J. Dundurs, Discussion of Edge-bonded dissimilar orthogonal elastic wedges under normal and shear loading, J. Appl. Mech. 36 (1969) 650.

[28] J.R. Rice, Elastic fracture mechanics concepts for interfacial cracks, J. Appl. Mech. 55 (1988) 98.

[29] C.F. Shih, R.J. Asaro, Elastic-plastic analysis of cracks on bimaterial interfaces: part I − small scale yielding, J. Appl. Mech. 55 (1988) 299.

[30] R.O. Ritchie, R.M. Cannon, B.J. Dalgleish, R.H. Dauskardt, J.M. McNaney, Mechanics and mechanisms of crack growth at or near ceramic-metal interfaces: interface engineering strategies for promoting toughness, Mater. Sci. Eng. A166 (1993) 221.

[31] F. Erdogan, G.C. Sih, On the crack extension in plates under plane loading and transverse shear, J. Basic Eng. 85 (1963) 519.

[32] C.H. Wu, Maximum-energy-release-rate criterion applied to a tension-compression specimen with crack, J. Elasticity 8 (1978) 235.

4

Nonlinear-elastic fracture mechanics (NLEFM)

4.1 Introduction

As we have described in Chapter 3, linear-elastic fracture mechanics (LEFM) provides a quantitative framework for the evaluation of the fracture toughness and for predicting failure loads in service for a broad class of materials. If one examines the toughness of various different materials as, for example, listed in Table 3.1 in Section 3.3.4, it is apparent that LEFM works well for high strength/low toughness materials, where the plastic zones are small enough in lab-sized samples to not exceed the criteria for small-scale yielding. This applies to virtually all ceramics, lower toughness polymers, lower toughness metallic alloys, and those with strengths above ~1000 MPa. However, it is not readily applicable for many lower strength structural alloys such as stainless steels, constructional pressure vessel, and piping steels where a tensile strength below ~1000 MPa is accepted so as to guarantee a high resistance to fracture.[1] The prime example of this is the nuclear pressure steel SA533B, which is a 500 MPa strength mild steel with a toughness exceeding 200 MPa√m. Resistance to fracture is clearly mandatory in this case for a component which houses the nuclear pile, particularly since the steel will progressively embrittle throughout its service life due to irradiation hardening. Valid K_{Ic} testing of this steel, however, involves prohibitively large samples (>0.6 m thick) that few laboratories can handle. Furthermore, because the steel is subjected to irradiation embrittlement, which is life-limiting, surveillance samples are housed in the reactor vessel to be

---

[1] As plastic deformation provides the main (intrinsic) contribution to the fracture toughness of nominally ductile materials, the properties of strength and toughness are often mutually exclusive [1].

49

extracted periodically during its life for testing to verify predictions for the degradation in the steel's toughness. For cost and space reasons, such surveillance samples by necessity must be small, which made the development of a new methodology to reliably measure the toughness of these lower strength/higher toughness materials an imperative. This primarily provided the motivation, initially sponsored by the nuclear and power industries, to enable "small sample" toughness testing through the development of nonlinear-elastic fracture mechanics (NLEFM) as an engineering methodology to predict fracture in the presence of plastic deformation.

4.2 Stress analysis of cracks: HRR singularity and J-integral

Although preceded by critical studies by Eshelby [2], Sanders [3], and Cherepanov [4],[2] the fundamentals of NLEFM, as we know it today, began with the work of Rice and Hutchinson in the mid- to late 1960s [5–7].[3] This work centered around the use of Rice's J-contour integral [7] to characterize crack-tip stress and displacement fields in the presence of plasticity to form the basis of an elastic-plastic fracture criterion. However, rather than follow this chronological order of developments, we will describe the fundamentals of NLEFM by analogy to LEFM, starting with the characterizing parameter definition of J.

4.2.1 J as a characterizing parameter

As described in Chapter 3, the restrictions in the use of LEFM lie with the fact that the analysis of the crack-tip stress and strain fields, the Williams $1/\sqrt{r}$ singularity [9] (Eq. 3.1), is derived for a linear-elastic constitutive law, and thus the inevitable presence of local plasticity at the crack tip must be small compared to the specimen/component dimensions. A resolution of this problem is clearly to solve for these crack-tip stress and displacements fields but inside the plastic zone, which is

---

[2] Eshelby [2] in the 1950s developed the energy momentum tensor which was similar to Rice's J-integral [7]. In fact, the J-integral is identical in form to the static component of this energy momentum tensor, although Eshelby had not applied it as an expression for crack-driving force. Sanders [3] in 1960 proposed a path-independent I-integral as a criterion for crack extension. Cherepanov [4] has been credited with developing the J-integral independently of Rice in 1967.

[3] In 1961, Wells [8] in England proposed an alternative concept to describe fracture in the presence of plasticity, that of the crack-tip opening displacement, which we will later describe as a fracture criterion in Chapter 5.

what Hutchinson, Rice and Rosengren attempted to do with their derivation of the so-called *HRR singularity* [5,6]. This solution was formulated for a nonlinear-elastic material (Fig. 4.1), displaying a so-called Ramberg-Osgood (power-hardening) constitutive law:

$$\frac{\bar{\varepsilon}}{\varepsilon_o} = \alpha \left(\frac{\bar{\sigma}}{\sigma_o} \right)^N , \tag{4.1}$$

where, respectively, $\bar{\sigma}$ and $\bar{\varepsilon}$ are the equivalent (Mises) stress and plastic strain, and σ_o and ε_o are the reference (yield) stress and strain, N is the strain hardening exponent (which varies between 1 for a linear-elastic material and ∞ for a fully-plastic one), and α is a dimensionless constant of order unity.

As per the coordinate system in Fig. 3.1, the HRR singularity gives the local stresses σ_{ij}, strains ε_{ij}, and displacements u_i, at distance r, angle θ, ahead of the crack tip, with $r \rightarrow 0$ as [5,6]:

$$\sigma_{ij} \rightarrow \sigma_o \left[\frac{J}{\alpha\sigma_o\varepsilon_o I_N r} \right]^{\frac{1}{N+1}} \widetilde{\sigma}_{ij}(\theta, N) , \tag{4.2a}$$

$$\varepsilon_{ij} \rightarrow \alpha\varepsilon_o \left[\frac{J}{\alpha\sigma_o\varepsilon_o I_N r} \right]^{\frac{N}{N+1}} \widetilde{\varepsilon}_{ij}(\theta, N) , \tag{4.2b}$$

$$u_i \rightarrow \alpha\varepsilon_o \left[\frac{J}{\alpha\sigma_o\varepsilon_o I_N r} \right]^{\frac{N}{N+1}} r \, \widetilde{u}_i(\theta, N) , \tag{4.2c}$$

where $\widetilde{\sigma}_{ij}(\theta, N)$, $\widetilde{\varepsilon}_{ij}(\theta, N)$, and $\widetilde{u}_i(\theta, N)$ are simply the angular functions for, respectively, stress, strain, and displacement, which, unlike the

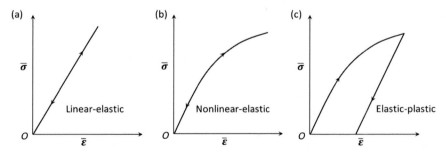

(a) $\bar{\sigma}$ Linear-elastic $\bar{\varepsilon}$

(b) $\bar{\sigma}$ Nonlinear-elastic $\bar{\varepsilon}$

(c) $\bar{\sigma}$ Elastic-plastic $\bar{\varepsilon}$

FIGURE 4.1   Schematic representations of the constitutive (stress-strain) loading and unloading behavior for (a) linear-elastic, (b) nonlinear-elastic, and (c) elastic-plastic solids. Although materials unload elastically after plastic deformation, the theory underlying J is based on the deformation theory of plasticity, which simplifies such elastic-plastic behavior as nonlinear-elastic behavior. This can still provide accurate solutions as long as the deformation is monotonic, increasing, and proportional. Unloading is thus not permitted.

Williams linear-elastic singularity in Eq. 3.1, are also a function of N, and I_N is an integration constant weakly dependent on r. Provided all higher order terms are ignored, the HRR field is controlled uniquely by J, thereby establishing J as the singular characterizing parameter for these crack-tip stresses and displacements determined for nonlinear-elastic deformation conditions.

4.2.2 J as a path-independent integral

So what exactly is J? J actually has three definitions. In addition to uniquely characterizing the HRR crack-tip fields, J was initially formulated as a path-independent integral [7]. For a static, homogeneous, isotropic solid displaying a nonlinear-elastic constitutive relationship (Fig. 4.1), if an arbitrary, anticlockwise path Γ is taken around the tip of a crack linking the lower to the upper crack surface, as depicted in Fig. 4.2, J can be defined as the line integral:

$$J = \int_{\Gamma} \left(W_{se} dy - T_i \frac{\partial u_i}{\partial x} ds \right) , \tag{4.3}$$

where ds is an increment of length on the contour Γ, T_i represents the n_j components of the traction vector, defining the stresses with respect to the outward unit vector n normal to the contour path, i.e., $T_i = \sigma_{ij} n_j$, u_i are the corresponding components of the displacement vector, and W_{se} is the strain energy density given by:

$$W_{se} = \int_0^{\varepsilon_{ij}} \sigma_{ij} \, d\varepsilon_{ij} . \tag{4.4}$$

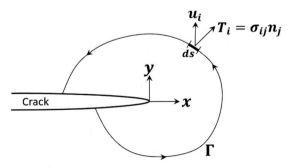

FIGURE 4.2   Schematic representation Rice's [7] anticlockwise contour Γ around a crack tip, connecting the lower to the upper surfaces of the crack, which is used to define the J-integral (Eq. 4.3). T_i is the traction vector, at length increment ds, representing the n_j components of the normal stresses acting on the boundary; u_i are the corresponding components of the displacement vector.

Rice [7] showed that the computed J-integral was independent of the contour path taken around the crack tip (hence the term path-independent), except in the crack blunting region in the immediate vicinity of the crack tip (note that for a totally closed contour, $J = 0$).

It is important to note here that the J-contour integral and the J-based HRR crack-tip fields are formulated using the *deformation theory of plasticity* which models elastic-plastic behavior in terms of nonlinear elasticity, i.e., where stresses are proportional to the current strains. As illustrated in Fig. 4.1, this can provide an accurate description of elastic-plastic behavior[4] provided the material is not unloaded, more specifically that the loading is monotonic, increasing, and proportional (i.e., the ratios of principal stresses remain constant during loading).[5] One thing to note here though is that akin to the specimen size requirements to use K to characterize fracture under SSY conditions, there are analogous validity criteria to ensure that the HRR field, characterized by J, actually does describe the relevant crack-tip stress and displacement distributions. These pertain to the regions of unloading and non-proportional loading that exist very close to the crack tip which need to be small enough to ignore. These requirements, however, are far less restrictive than the SSY requirements, as we discuss below and in the description of the standard solutions for J and the J_{Ic} fracture toughness test, respectively, in Sections 4.4 and 4.5.

4.2.3 J as an energy parameter

The third definition of J is as the strain energy release rate for a nonlinear-elastic solid containing a crack. Specifically, Rice demonstrated that the value of J was given by the rate of change in potential energy per unit increase in crack area for a nonlinear-elastic solid, i.e., it could be equated to G under linear-elastic conditions [7]:

$$J = - \frac{dU_{PE}}{dA} . \qquad (4.5)$$

For a cracked plate subject to an external force P (Fig. 4.3), similar to the description of G for linear-elastic conditions in Section 3.4, A is the crack area here and the potential energy, U_{PE}, is given in terms of the strain energy, U_ε, and the work done, W_D, by the external forces, viz:

$$U_{PE} = U_\varepsilon - W_D . \qquad (4.6)$$

---

[4] Elastic-plastic behavior, e.g., in Fig. 4.1c, can be modeled using the more complex J_2 *flow theory of plasticity*, where the total stresses are more accurately related to the incremental strains.

[5] As Eshelby once remarked: "…but the material may call your bluff!"

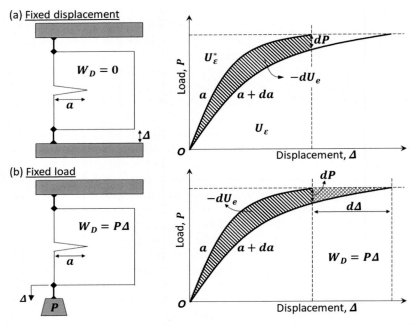

FIGURE 4.3   A nonlinear-elastic plate containing a crack of length a "loaded" (a) at a fixed displacement Δ, and (b) at a fixed load P, relevant to the energy definition of J, in Eqs. (4.5)–(4.8).

As the work done under displacement control is zero, whereas under load control, $W_D = P\Delta$, the potential energy will be given by:

$$U_{\mathrm{PE}} = U_\varepsilon \, , \quad \text{(displacement control)} \tag{4.7a}$$

$$U_{\mathrm{PE}} = U_\varepsilon - P\Delta = U_\varepsilon^* \, , \quad \text{(load control)} \tag{4.7b}$$

where $U_\varepsilon^* = \int_0^P \Delta dP$ and is termed the complimentary strain energy. Accordingly, from Eq. (4.5), J can be expressed as:

$$J = \left(\frac{dU_\varepsilon}{dA}\right)_\Delta = -\int_0^\Delta \left(\frac{\partial P}{\partial A}\right)_\Delta d\Delta \, , \quad \text{(displacement control)} \tag{4.8a}$$

$$J = \left(\frac{dU_\varepsilon^*}{dA}\right)_P = \int_0^P \left(\frac{\partial \Delta}{\partial A}\right)_P dP \, . \quad \text{(load control)} \tag{4.8b}$$

As with the linear-elastic solutions for $\mathcal{G}$ in Eqs. 3.10–3.12, these nonlinear-elastic solutions for J under displacement vs. load control are the same, differing only in sign and the infinitesimal amount of $\frac{1}{2}dPd\Delta$.

It is apparent from these derivations, the strain energy release rate $\mathcal{G}$ is a special case of *J* under linear-elastic conditions. Thus, for a linear-elastic solid under mode I loading:

$$J = \mathcal{G} = \frac{K_I^2}{E'} . \tag{4.9}$$

4.3 *J* as a fracture criterion

4.3.1 $J = J_c$

As described in the previous section, it is apparent that Rice's *J*-integral is both the singular characterizing parameter, akin to *K*, of the HRR singularity for the stresses and displacements in the vicinity of a crack tip, and an energy parameter, akin to $\mathcal{G}$, in defining the strain energy release rate, in both cases for a nonlinear-elastic solid. In light of this, it would seem more than feasible that *J* could be utilized as a criterion for fracture in the presence of plasticity. The presumption is that provided the HRR field (Eq. 4.2) exists at the tip of a nominally atomically sharp crack, such that *J* singularly characterizes the local stresses and displacements over the dimensions where the salient fracture events occur, then the onset of fracture should occur at a critical value of *J*, i.e., when $J = J_c$ (or J_{Ic} in plane-strain). This was the question that Begley and Landes set out to demonstrate in the early 1970s [10,11].

The issue at stake then was that although constrained geometries, like the edge-cracked bend configuration and the compact-tension sample, have the same $1/\sqrt{r}$ *K*-fields describing the crack-tip stress and strain fields for linear-elastic conditions as the relatively unconstrained middle-cracked plate geometry, i.e., the *K*-based stress fields are unique, under fully plastic conditions defined by plane-strain slip-line field theory, this simply is not the case. Indeed, as McClintock [12] pointed out, there is nothing unique about these fully plastic crack-tip fields; the slip-line fields for a fully yielded double-edge notched plate in tension and single-edge notched plate in bending, which essentially develop the so-called Prandtl field,[6] have fundamentally different near-tip stress and strain fields compared to the middle-cracked plate in tension (Fig. 4.4). The Prandtl field develops high triaxiality with a maximum σ_{yy} stress ahead of the

---

[6] Finite element solutions for SSY/plane-strain conditions in an elastic-perfectly plastic solid show that the full Prandtl field is developed ahead of the crack tip for a double edge-cracked plate in tension, and almost developed in the edge-cracked plate in bending; however, the corresponding crack-tip stress field for a fully yielded middle-cracked plate in tension is completely different [13].

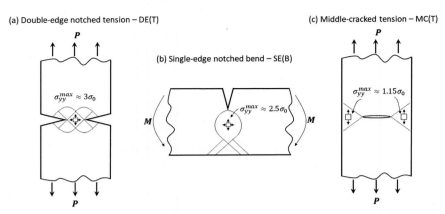

FIGURE 4.4   Slip-line fields for the fully yielded configurations of (a) double-edge notched tension − DE(T), (b) single-edge notched bend − SE(B), and (c) middle-cracked tension − MC(T), showing the non-uniqueness of the crack-tip stress fields based on rigid perfectly plastic (i.e., non-hardening), plane-strain, deformation.

crack tip of roughly 2.5 σ_o, where σ_o is the yield or flow stress, whereas for the middle-cracked plate in tension, only modest triaxiality is attained ahead of the crack tip, with the maximum $\sigma_{yy} \sim 1.15\, \sigma_o$, but intense shear strains are created on planes at 45 degrees to the crack.

However, such slip-line field solutions pertain to rigid perfectly plastic, plane-strain deformation, and rationalizing such non-unique, fully plastic crack-tip field solutions with the concept of J being a single-valued, configuration-independent descriptor of the crack-tip stress and displacement fields requires that some strain hardening must exist. Accordingly, Begley and Landes attempted to measure the plane-strain fracture toughness in terms of J_{Ic}, on the presumption that although the fully-plastic, non-hardening crack-tip stress and displacements fields were not the same for the different cracked geometries, the inevitable presence of strain hardening in real metallic materials would confer some uniqueness at the crack tip in terms of the existence of an HRR field [11].[7] To verify this, they tested under-sized, yet plane-strain, samples of an

---

[7] It is interesting to note here that for materials that show little to no strain hardening, such as bulk-metallic glasses, although K_{Ic} measurements in these amorphous alloys are independent of specimen size provided SSY conditions prevail, higher toughnesses measured in the J-regime are distinctly variable and size-dependent [14], even though they are valid in terms of the ASTM E1820 Standard [15]. The premise here is that due to the lack of strain hardening, the existence of the HRR field in these materials is highly limited. As a general rule for non-hardening materials in the presence of a significant degree of plasticity, a single-parameter description of the fracture toughness will not be valid and, as such, will be size- and geometry-dependent.

intermediate strength NiCoMoV steel in the edge-cracked bending and middle-cracked tension geometries, where the fully-yielded, non-hardening fields are totally different (Fig. 4.4), and measured a constant J_{Ic} at crack initiation in both geometries, that was the same as the G_{Ic} measured in a larger-sized sample meeting the SSY size requirements (Eq. 3.5). *The $J = J_c$ fracture criterion, for measuring the fracture toughness in samples not large enough to meet the highly restrictive $K = K_c$ (or $G = G_c$) SSY conditions, was born!*

The only problem was that the authors had made a mistake in the calculation of the toughness for their middle-cracked tension samples. Although they claimed that they had made two errors that were self-cancelling (they also incorrectly assumed that crack initiation occurred at maximum load) [16], this was an unfortunate situation that was not resolved until some 5 years later when Parks and McMeeking [17] performed full numerical calculations of the stress and displacement fields of both cracked configurations and compared them with the corresponding fields based on *J*. They found that there was indeed uniqueness in the crack-tip fields for these two different geometries, as Begley and Landes had postulated, but that the sample size requirements for *J*-dominance at the crack tip for the middle-cracked tension configuration were at least an order of magnitude larger than for the constrained Prandtl-field geometries such as the edge-cracked bend. Accordingly, unlike the sample validity criteria for linear-elastic K_c and K_{Ic} testing in Eq. 3.5, the corresponding size requirements for J_c and J_{Ic} can be far less restrictive *but are a function of sample geometry*. It is for this reason that the middle-cracked tension — MC(T) geometry is not a specified test geometry in the ASTM Standard E1820 [15] procedures for measuring the J_{Ic} fracture toughness. These procedures are described in Section 4.5.

If we confine our attention to the Prandtl-field-like crack configurations that are specified in this Standard, such as the edge-cracked bend and compact-tension, as noted above there are sample size requirements that need to be met to confirm that the HRR field, and hence *J*, actually describe the local crack-tip stresses and displacements over relevant dimensions ahead of the tip. Just as the plastic zone is a violation of the linear-elastic *K*-field, which sets the SSY size requirements (Eq. 3.5), there are regions at the crack tip where unloading will occur (where $r \sim \Delta a$) and a somewhat larger region of non-proportional plastic loading, both of which violate the initial premise of nonlinear elasticity in postulating the HRR field; both regions need to be small enough to ignore. The ASTM E1820 Standard [15] has relaxed its requirements somewhat on this issue but now specifies that for J-dominance, the remaining uncracked ligament b must be large enough to satisfy:

$$b \geq \frac{10\,J}{\sigma_o}\,, \quad \text{(for J-dominance)} \tag{4.10a}$$

where σ_o is the effective yield stress (or flow stress) equal to the average of the yield, σ_y, and ultimate tensile stress, σ_{UTS}: $\sigma_o = \frac{1}{2}(\sigma_y + \sigma_{UTS})$.[8] Additionally, for plane strain, the thickness, B, should exceed:

$$B \geq b\,. \quad \text{(for plane strain)} \tag{4.10b}$$

You will notice that these size requirements for performing nonlinear-elastic fracture toughness testing (in Eq. 4.10) are far less restrictive than those for performing LEFM testing. If we consider the SA533B nuclear pressure vessel steel again, where it would take a sample with an un-cracked ligament and specimen thickness exceeding 0.6 m to measure K_{Ic} (Table 3.1), the corresponding critical sample dimensions to measure J_{Ic} from Eq. (4.10) are reduced to 5 mm.[9] Moreover, because of the equivalence of J and $\mathcal{G}$ under linear-elastic conditions:

$$J_{Ic} = \mathcal{G}_{Ic} = \frac{K_{JIc}^2}{E'}\,, \quad \text{(under linear-elastic conditions)} \tag{4.11}$$

the K_{Ic} value can be back-calculated from the small specimen J_{Ic} measurement (this is termed K_{JIc}) to represent the fracture toughness value if the full 0.6 m-sized specimen had been tested.

Thus, the HRR field firmly establishes J as the unique and singular characterizing parameter that controls the crack-tip stresses and strains in a plastically deforming body (without unloading). As long as the crack-tip region is represented by this HRR field over the scale of fracture events, then a critical value of J can be used as a fracture criterion to represent the crack initiation toughness. However, there are still clearly differences in constraint that can exist between different test geometries which can affect the precise magnitudes of these crack-tip fields. This has led to an interest in developing two-parameter fracture criteria. We briefly discuss this in Section 4.7 on the T-stress.

---

[8] It has been estimated that the corresponding size requirement for a valid J_{Ic} measurement in the middle-crack tension geometry may exceed $200\,J/\sigma_{flow}$ for materials with moderately low strain hardening (e.g., for N ~ 10).

[9] In this calculation for the SA533B steel, $J_{Ic} = 263$ kJ/m^2 ($K_{Ic} = 245$ MPa$\sqrt{m}$) and $\sigma_o = 538$ MPa ($\sigma_y = 500$ MPa).

4.3.2 Importance of validity criteria (size requirements)

Fracture mechanics is somewhat unusual with respect to other forms of engineering mechanics in that all the relationships that we use for fracture criteria, be them based on K, J, or other characterizing parameters, are subject to validity criteria, in particular which specify the size requirements for the test samples to be used in the measurements. We have noticed that today, many publications find their way into the technical literature where such criteria are either ignored or marginalized, to the effect that the resulting measured K_{Ic} or J_{Ic} fracture toughness values are size- and geometry-dependent, or at worst completely wrong. The validity criteria are literally mandatory to take in account, whether one is dealing with limitations to the size of the plastic zone with the use of the stress intensity K, or the extent of the unloading and non-proportional loading zones with the corresponding use of J, in both cases to characterize the relevant crack-tip fields involved to accurately measure the toughness. This is succinctly demonstrated by an illustration in Anderson's textbook [18], reproduced in Fig. 4.5, which demonstrates how excessive local plasticity can completely invalidate the use of K- or J-based crack-tip fields. This essentially asks the question, ... if I am going to use K to measure the fracture toughness, does K actually characterize the local stresses and strains that are involved in the fracture events?... and, if there is too much local plasticity in the crack-tip region, does the approximation of the nonlinear-elastic HRR field mean that J better describes, indeed actually describes, the local stresses and strains where the critical fracture events are taking place, to justify a $J = J_c$ fracture criterion? ... or are *all bets are off*? When SSY size requirements are met, i.e., $r_y << a$, b, from Fig. 4.5a the actual local crack-tip fields are characterized by the K-field so that the use of $K = K_c$ is realistic. When the extent of local plasticity is too large (Fig. 4.5b), the LEFM K-field singularity no longer characterizes any region of the actual crack-tip stress fields, but regions within the plastic zone can still be described by the HRR singularity and so the use of a $J = J_c$ criterion is justified. With larger scale plasticity, we have no singular parameter characterization of the crack-tip stresses and strains, and so K- and J-based fracture mechanics methodologies are simply not applicable.

4.4 *J*-solutions

Unlike solutions for the stress intensity factors (Box 3.2), which can be found in numerous books and publications for various cracked configurations and test geometries, corresponding tabulations of solutions for J are far less available. J-solutions for test geometries, e.g., specified in the ASTM Standard E1820 [15] for fracture toughness measurements, are available in that Standard and in a few textbooks [18,19]. But for a more

4. Nonlinear-elastic fracture mechanics (NLEFM)

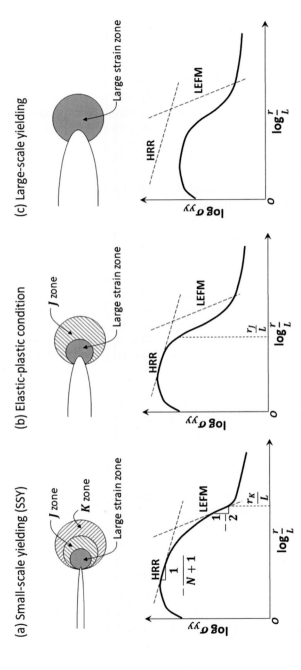

FIGURE 4.5 Importance of validity requirements in maintaining the appropriate crack-tip stress and displacement fields in the presence of plasticity in order to justify the valid use of a critical value of K or J as a criterion for fracture. r/L is the ratio of the distance ahead of the crack tip, normalized by the size of the structure or specimen L; N is the strain hardening exponent from the constitutive law in Eq. (4.1) [18]. *Adapted from T. L. Anderson, Fracture Mechanics: Fundamentals and Applications, fourth ed., CRC Press, Boca Raton, FL (2017). Copyright 2017. Reproduced by permission of Taylor and Francis Group, LLC, a division of Informa plc.*

extensive set, one needs to access a report published in 1981 by the Electric Power Research Institute in Palo Alto, CA, by V. Kumar et al. [20], which provides a comprehensive compendium of solutions that could also be applied to structures in service. Below we reproduce a few *J*-solutions for standard test geometries.

4.4.1 Deep single-edge cracked bend specimen

One of the first *J*-solutions was developed by Rice for the deep single-edge cracked bend sample [21]; indeed, this relatively simple derivation served as the basis for several other test geometries. For the sample shown in Fig. 4.6, where deeply-cracked means an $a/W > {\sim}0.6$, the application of the moment M causes an angular rotation ϕ, which is comprised of the angle ϕ_{uc} if the plate was uncracked, plus that due to the presence of the crack ϕ_c, where $\phi_c \gg \phi_{uc}$. Using dimensional analysis, Rice showed that provided the material properties were constant, the angle ϕ_c would only be a function of the bending moment M, the uncracked ligament b (as the plate was deeply cracked), and of course its thickness B:

$$\phi_c = f\left(\frac{M}{Bb^2}\right). \tag{4.12}$$

If an experimental M vs. ϕ plot is obtained (as in Fig. 4.6), as ϕ_{uc} does not vary with a, using Eq. (4.8), J can be expressed as:

$$J = \frac{1}{B}\int_0^M \left(\frac{\partial \phi_c}{\partial a}\right)_m dM = -\frac{1}{B}\int_0^M \left(\frac{\partial \phi_c}{\partial b}\right)_m dM. \tag{4.13}$$

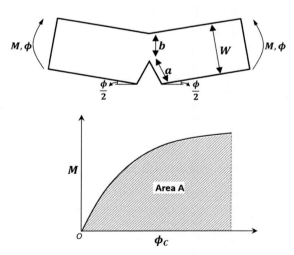

FIGURE 4.6 Deep single-edge cracked bend specimen used in the calculation of the *J*-solution, showing the experimentally measured moment M vs. angle $\phi \approx \phi_c$ curve.

Differentiating Eq. (4.12) with respect to b and incorporating in Eq. (4.13) gives an expression for J of:

$$J = \frac{2}{Bb} \int_0^{\phi_c} M d\phi_c = \frac{2}{Bb} A, \qquad (4.14a)$$

where $A = \int_0^{\phi_c} M d\phi_c$ is the area under the M vs. ϕ curve in Fig. 4.6 (as for a deeply-cracked sample, ϕ_{uc} is small enough to ignore). The ASTM Standard calls for a three-point bend test. If P is the point load (as per the configuration shown in Fig. 3.3) and Δ is its displacement, then J-solution for the deep edge-cracked bend specimen can be expressed as:

$$J = \frac{2}{Bb} \int_0^{\Delta} P d\Delta = \frac{2}{Bb} A, \qquad (4.14b)$$

where A is the absorbed energy represented by the area under the load-displacement curve.

4.4.2 General form of J-solution for various specimen geometries

Eq. (4.14) forms the basis of the J-solutions for many specimen geometries, which are generally expressed in terms of the elastic and plastic contributions to J as:

$$J = J_{el} + J_{pl} = \frac{\eta A_{el}}{Bb} + \frac{\eta A_{pl}}{Bb}, \qquad (4.15a)$$

where A_{el} is the recoverable elastic energy, A_{pl} is the plastic energy absorbed by the specimen, and η is a dimensionless constant which is a function of specimen geometry. This relationship is generally expressed in the form:

$$J = \frac{K_I^2}{E'} + \frac{\eta A_{pl}}{Bb}. \qquad (4.15b)$$

As derived above, $\eta = 2$ for the deep edge-cracked bend SE(B) sample. For the compact-tension C(T) sample, $\eta = 2 + 0.522\, b/W$ [22].

4.5 Measurement of the fracture toughness J_{Ic}

Similar to the measurement of K_{Ic}, outlined in Section 3.3.3, procedures for conducting J_{Ic} fracture toughness tests are described in detail in ASTM Standard E1820 [15]. The intent is the same: to determine the driving force to cause crack initiation from a pre-existing (near-atomically sharp) crack, i.e., a fatigue crack, in a test specimen that satisfies the conditions that, in this case the HRR field and hence J, define the distribution of the local stresses and displacements in the vicinity of this pre-crack. As you might expect, there are several important differences to the K_{Ic} procedures though, which principally pertain to (i) the choice of specimen geometry, (ii) the requirement to measure *load-line* displacements, (iii) the need to monitor crack extension to determine the R-curve, and (iv) of course, the nature of the validity criteria.

First, with respect to an appropriate test specimen, the Standard excludes the middle-cracked tension geometry due to the reasons discussed in Section 4.3, and prescribes the Prandtl-field-like geometries including the single-edge notched bend sample, tested in three-point bending, and the various forms of compact-tension specimens; these samples need to have somewhat longer pre-cracks ($0.5 \leq a/W \leq 0.7$) than for K_{Ic} testing to enhance the sensitivity of compliance measurements that are often used to monitor crack extension. An additional important requirement is that the specimen must be designed to enable measurement of the load-line displacement; unlike K_{Ic} measurements where the monitored displacement merely needs to be linear, with J-based testing we are measuring an energy which requires information on the load and the actual displacement of that load. A typical design of a compact-tension specimen to enable such measurements is shown in Fig. 4.7. The throat of the notch is opened to permit an extensometer, e.g., a clip or displacement gauge, to directly measure the displacement along the line of the application of the load P, i.e., the so-called load-line displacement, Δ_{LL}.

Because of the more extensive plasticity associated with J-based testing, which stabilizes stable crack extension, the load-displacement curves are distinctly nonlinear. Accordingly, since J_{Ic} is presumed to be measured at the onset of crack initiation, this has to be independently

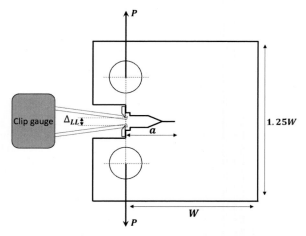

FIGURE 4.7 A compact-tension C(T) specimen designed to permit load-line displacement measurements required for the measurement of the J_{Ic} fracture toughness.

detected. One method, albeit rather tedious, is to utilize multiple identical specimens, loading them to different levels, before unloading and measuring the amount of stable crack extension.[10] A more expedient method though is to use a single specimen where the crack extension is monitored in situ. There are various techniques for doing this, such as the electrical potential method [23], but the most used technique is to utilize unloading compliance measurements (Fig. 4.8), using a clip gauge or back-face strain gauge [18]. Since the material will unload elastically, by periodically unloading <20% of the load throughout the test and using the relevant compliance calibrations (many of which are listed in the Standard), the amount of stable crack extension, Δa, at each load level can be determined. Coupled with the values of J computed for the pertinent test geometry, e.g., of the form of Eq. (4.15), viz:

$$J = \frac{K_I^2}{E'} + \frac{\eta A_{pl}}{B_N b_o} . \tag{4.16}$$

where A_{pl} is the plastic (non-recoverable) area under the load vs. load-line displacement curve (Fig. 4.8), B_N is the net specimen thickness equal to B or as defined in Fig. 3.10 for a side-grooved specimen, b_o is the initial uncracked ligament depth, given by $(W - a_o)$, and η is the dimensionless constant for the particular geometry, e.g., equal to 2 for the SE(B) sample and $2 + 0.522\, b_o/W$ for the C(T) sample. Based on these measurements, a

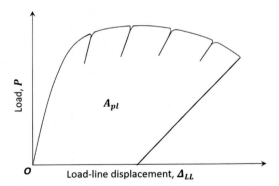

FIGURE 4.8 Load P vs. load-line displacement Δ_{LL} curve during J_{Ic} measurement showing the use of periodic unloading to estimate the crack length from the compliance measured by the reciprocal of the unloading slope. A_{pl} is the plastic area under the curve, shown here for the final point on the curve; this the plastic energy absorbed by the sample up to that point.

[10] Heat tinting, by running a flame over the specimen to oxidize the fracture surface where the slow crack growth had occurred, prior to breaking open the specimen in liquid nitrogen, can often help in identifying the region of stable crack extension, Δa.

J_R vs. Δa R-curve[11] (Fig. 4.9) is constructed from which the J_{Ic} can be determined.

The determination of J_{Ic}, which is measured at the point of crack initiation, in essence involves extrapolating the $J_R(\Delta a)$ R-curve to the value of J where $\Delta a \rightarrow 0$; however, the ASTM E1820 [15] procedures are somewhat more complicated than that, as illustrated in Fig. 4.9. One does extrapolate the R-curve but only using data between two exclusion lines, located at Δa values of 0.15 and 0.5 mm, with a slope of $2\sigma_0$, where σ_0 is the flow stress (or effective yield stress) given by the average of the yield and

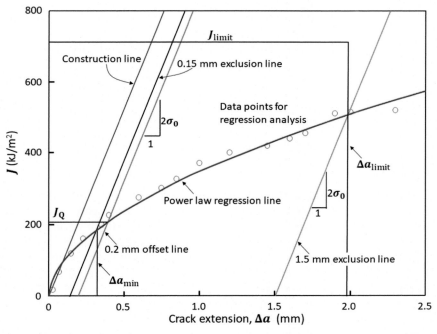

FIGURE 4.9 The $J_R(\Delta a)$ resistance curve for the measurement of the provisional crack initiation fracture toughness, J_Q. The value of J_Q can be identified with J_{Ic} provided the validity criteria, in terms of specimen size requirements, meet the requirement of J-dominance and plane-strain conditions [15] (reproduced with permission).

[11] It should be noted here that the crack size, actually the remaining ligament size b, is not updated for crack extension along the R-curve. This is because that the R-curve here is used specifically to measure the J_{Ic} value at crack initiation, which would be unaffected by any subsequent stable crack growth. However, as we discuss in Section 4.6, if determining the crack-growth toughness is also the objective of the measurements, which is a function of the full R-curve, the values of b used in the computation of J must be updated for the occurrence of stable crack advance.

ultimate tensile stresses. Additionally, there are two more exclusion lines which specify the maximum crack extension (at $\Delta a = 2$ mm) and the maximum J ($J_{max} = b_o \sigma_o / 15$) that can be measured with this specimen. The data on the R-curve within these exclusion limits are fit to a power law:

$$J = C_1 (\Delta a)^{C_2}, \tag{4.17}$$

and where this power-law line intersects a 0.2 mm offset (blunting) line (again with a slope of $2\sigma_o$)[12] sets the provisional value of J_Q. The value of J_Q can then be identified with J_c provided it meets the validity criteria in terms of specimen size requirements for J-dominance:

$$b_o \geq 10 \frac{J_Q}{\sigma_o}, \tag{4.18a}$$

and with J_{Ic} if the conditions for plane-strain are additionally met:

$$B \geq 10 \frac{J_Q}{\sigma_o}. \tag{4.18b}$$

4.6 $J_R(\Delta a)$ resistance curves

4.6.1 Crack-growth toughness

For structural materials, it is invariably a requirement by designers that they possess some degree of tensile ductility. What this means in terms of fracture is that they exhibit a rising R-curve, i.e., they display some stable cracking prior to failure, rather than fracturing catastrophically without warning. As plastic deformation is the principal factor that promotes such stable crack growth prior to fracture instability, the $J_R(\Delta a)$ resistance curve becomes an important "material property" to evaluate toughness. We are all initially taught that toughness is a combination of strength and ductility (the chemists still measure it that way!), and to the first approximation, of course it is. But certain materials can possess strength and ductility yet still be susceptible to premature failure from the inevitable presence of cracks and flaws. Accordingly, we firmly believe that the measurement of the J-based R-curve, which incorporates both the contribution to toughness from plasticity and the "crack-growth toughness" associated with the stable propagation of cracks prior to failure, is essential to characterizing the damage-tolerance of any potential structural material.

---

[12] The 0.2 mm offset line and the 0.2 and 2 mm exclusion lines are defined by the expression $J = 2\sigma_o \Delta a$ in an attempt to estimate, and hence separate out, the apparent crack extension due to crack-tip blunting.

Whereas this approach is perfectly acceptable from a materials science perspective, it does present difficult problems from a mechanics perspective, because unlike K_{Ic} and J_{Ic} which can be defined to be material properties, the R-curve can be a function of geometry, as we discussed in Section 3.5 for small-scale yielding conditions. Furthermore, the situation is even less robust under nonlinear-elastic conditions, in part due to the limited HRR field at the tips of cracks in non-Prandtl-field geometries, such as the MC(T), and perhaps more importantly because the notion of J-controlled crack growth is somewhat of an oxymoron. Not only does the material ahead of the crack tip unload with crack advance, but a prime factor responsible for stabilizing crack growth is the trail of plasticity in the wake of the crack tip as it advances (Fig. 4.10), and neither of these phenomena are consistent with the notion of modeling plastic deformation by nonlinear elasticity, i.e., the singular J-field will cease to exist with extensive crack advance and plasticity.

Notwithstanding these complexities in defining the precise crack-tip fields for growing cracks, we now examine the $J_R(\Delta a)$ resistance curves from the perspective of characterizing the crack-growth toughness.

A schematic of a $J_R(\Delta a)$ R-curve, shown in Fig. 4.11, indicates the various regimes of behavior with crack extension. The crack initially experiences crack-tip blunting (this results in apparent crack extension which is corrected through the use of the offset blunting line, as described in Section 4.5) prior to the onset of stable cracking where J_{Ic} is measured; this defines the crack-initiation fracture toughness. The slope of the subsequent R-curve, dJ_R/da, can be used as a measure of the crack-growth fracture toughness; alternative parameters that have been used for the crack-growth toughness are the J value at the limit of Δa_{max} or, if the specimen is large enough, the steady-state value, J_{ss}.

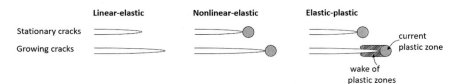

FIGURE 4.10 Idealization of crack extension for (a) a linear-elastic crack, where deformation in any crack-tip zone is fully recoverable on unloading due to crack advance, (b) a nonlinear-elastic crack, where the crack-tip plastic zone is modeled in terms of nonlinear elasticity and therefore the deformation is also fully recoverable, and (c) an elastic-plastic crack, which realistically models the plastic zone to unload elastically with crack advance and thus to leave a trail of plastically deformed material in the wake of the crack.

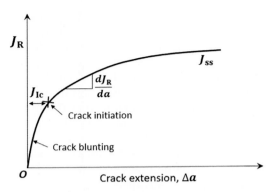

FIGURE 4.11 Schematic of a $J_R(\Delta a)$ crack resistance curve used to measure the crack growth toughness. Whereas J_{Ic} is used to measure the crack initiation toughness, the crack growth toughness can be defined in terms of the slope of the R-curve, e.g., dJ_R/da, or by a value of J at a specific crack extension Δa_{max} or at steady-state, J_{ss}.

The slope of the R-curve, dJ_R/da, is sometimes described in terms of the so-called *tearing modulus*, T_R, where dJ_R/da is nominalized with respect to the flow stress σ_o and Young's modulus E, as [24]:

$$T_R = \frac{E}{\sigma_o^2} \frac{dJ_R}{da}. \tag{4.19}$$

This dimensionless parameter is another measure of the crack-growth toughness.

Although these crack-growth toughness parameters can be difficult to define uniquely and are sometimes used rather arbitrarily, they are important for tough, ductile materials because, compared to the crack-initiation toughness, (i) most of the fracture toughness is represented by the crack-growth resistance on the R-curve, and (ii) microstructural effects invariably have a far greater influence on this crack-growth toughness. The toughest materials tend to have steeply rising R-curves with significant stable crack extension, both of which are indicative of damage-tolerant materials that are resistant to catastrophic fracture.

4.6.2 Measurement of $J_R(\Delta a)$ resistance curves

The measurement of the $J_R(\Delta a)$ resistance curve for the purpose of determining the complete R-curve and the crack-growth toughness follows the procedures [15] described in Section 4.5 for the measurement of J_{Ic}. However, there are a few important differences. Firstly, single specimen testing is best used for this purpose, with a crack monitoring technique employed, such as unloading compliance (Fig. 4.8), to keep track of the crack length a, and hence uncracked ligament b, used in the calculation of J. Secondly, as the crack is growing, the value of J at each point must be calculated incrementally. Recall that all this was unnecessary in the construction of the R-curve in Fig. 4.9 as it would have little to no effect on the value of J at the onset of crack extension. ASTM Standard E1820 lays out the appropriate procedures, as detailed below.

For each crack length data point, a_i, the J_i-integral is computed as the sum of elastic, $J_{el(i)}$, and plastic components, $J_{pl(i)}$, as indicated in Section 4.4.2:

$$J_i = K_i^2/E' + J_{pl(i)} \,, \tag{4.20}$$

where K_i is stress intensity corresponding to each data point on the load − load-line displacement curve. The plastic component of J_i can be calculated from the following equation:

$$J_{pl(i)} = \left[J_{pl(i-1)} + \left(\frac{\eta_{pl(i-1)}}{b_{(i-1)}} \right) \frac{A_{pl(i)} - A_{pl(i-1)}}{B_N} \right] \left[1 - \gamma_{(i-1)} \left(\frac{a_{(i)} - a_{(i-1)}}{b_{(i-1)}} \right) \right] \,, \tag{4.21}$$

where B_N is the net section thickness, $\eta_{pl(i-1)} = 2$ for the single-edge notched bend − SE(B) specimen and $2 + 0.522\, b_{(i-1)}/W$ for the compact-tension C(T) specimen, and $\gamma_{pl(i-1)} = 1$ for the SE (B) specimen and $1 + 0.76\, b_{(i-1)}/W$ for the C(T) specimen. $A_{pl(i)} - A_{pl(i-1)}$ is the increment of plastic area underneath the load-displacement curve (Fig. 4.8), and b_i is the uncracked ligament (i.e., $b_i = W - a_i$). Using this formulation, the value of J_i can be determined at any point along the load-displacement curve and together with the corresponding crack lengths, the $J_R(\Delta a)$ resistance curve can be created. Note that here, Δa is the difference in the individual crack lengths, a_i, during testing and the initial crack length, a after pre-cracking.

The ASTM Standard E1820 [15] specifies validity criteria for these R-curve measurements in terms of specimen size requirements for both crack-tip J-dominance, plane-strain conditions, and the extent of stable crack extension, namely:

$$B, b_o \geq \frac{10 J_{max}}{\sigma_o} \,, \tag{4.22a}$$

$$\Delta a_{max} \leq 0.25\, b_o \,. \tag{4.22b}$$

4.6.3 Measurement of J-toughness values for cleavage fracture

The above procedures for measuring the $J_R(\Delta a)$ resistance to determine J_{Ic} and the crack-growth toughness pertain largely to fracture by ductile fracture, i.e., microvoid coalescence. However, many lower- to medium-strength steels will display ductile stable cracking in the presence of significant plastic deformation yet still eventually fail catastrophically by brittle cleavage fracture. Because such cleavage fracture is promoted by high triaxial stresses, ASTM in their E1820 Standard [15] have specified a separate set of validity criteria for the value of J, termed J_c, at the onset of catastrophic brittle fracture following substantial stable cracking, specifically that the specimen size requirements meet:

$$B, \, b_0 \geq \frac{100J_c}{\sigma_0} \, , \qquad (4.23a)$$

$$\Delta a_{max} \, \leq \, 0.20 \text{ mm} + \frac{J_c}{2\sigma_0} \, . \qquad (4.23b)$$

4.7 T-stress and the modification of crack-tip fields

We have described how the Williams $1/\sqrt{r}$ linear-elastic stress and strain fields ahead of a crack tip (Eq. 3.1) and the corresponding HRR nonlinear-elastic fields (Eq. 4.2) can be uniquely characterized by a single parameter, respectively, the stress intensity K and J-integral, and as such how we can utilize them as size- and geometry-independent terms to characterize both stable and unstable fracture. However, these assertions are based on the fact that K and J both characterize the amplitude of the first term of their respective expansion series describing the crack-tip stresses and strains, and further that all the higher order nonsingular terms are small enough to be ignored. Despite this, as we have shown for the K-based singularity in Fig. 3.13, the deviation from the K-field (first term) description of the stresses from the complete (all terms) description will increase with distance r ahead of the crack tip and with geometry, i.e., these higher order terms are size- and geometry-dependent. This is even more important with nonlinear-elastic fracture mechanics where with increasing r, the crack-tip stresses can deviate markedly from the J-field, especially in certain geometries such as the middle-cracked tension. Indeed, in the presence of significant crack-tip plasticity, the active damage and fracture processes are much more likely to occur over significant dimensions ahead of the crack tip.

In this section, we will examine the effect of the first of these higher order terms, the so-called elastic T-stress for the linear-elastic crack-tip stress field, which actually does not vanish as $r \rightarrow 0$. We will also mention, but not in too much detail, the corresponding approach for nonlinear-elastic behavior, known as Q, which provides the basis for the two-parameter J-Q approach for characterizing the toughness.

4.7.1 Definition of the T-stress

If we re-examine the Williams singularity [9] for the linear-elastic crack-tip fields in an isotropic solid in mode I (Eq. 3.1), but now consider more than just the first term in the expansion series, then as $r \to 0$:

$$\sigma_{ij} \to \frac{K_I}{\sqrt{2\pi r}} f_{ij}(\theta) + \begin{bmatrix} T & 0 & 0 \\ 0 & 0 & 0 \\ 0 & 0 & \nu T \end{bmatrix} . \tag{4.24}$$

There is a second term, termed the elastic T-stress, which is also finite at the crack tip and represents a constant uniform stress parallel to the plane of the crack in the x-direction (with a corresponding νT in the z-direction in plane-strain, where ν is Poisson's ratio). The magnitude of T is proportional to the nominal stress in the vicinity of the crack:

$$T = \beta' \frac{K_I}{\sqrt{\pi a}} = \beta' Y \sigma^\infty , \tag{4.25}$$

where β' is the *biaxiality ratio* relating T to K_I ($\beta' = T\sqrt{\pi a}/K_I$) [25] and K_I has the usual definition in terms of the applied stress σ^∞, crack size a, and geometry factor Y (Eq. 3.2). T was first recognized by Larsson and Carlsson [26] to account for much of the geometry-dependent differences in the crack-tip fields. T also has a significant effect on the size and shape of the plastic zone, and as a function of Y, varies with a/W [27]. Solutions for the T-stress for a variety of two- and three-dimensional cracked geometries, e.g., the C(T), SE(B), MC(T), etc., can be found in Ref. [28].

So what is the effect of the T? Actually, T can be either negative or positive. Negative values of the T-stress, which represent a uniform compressive stress, act to lower the crack-tip constraint and hence reduce the stress intensification at crack tip, which has the effect of increasing the apparent fracture toughness. The $T = 0$ case only strictly applies to the limit of small-scale yielding where the K- or J-dominant singular term uniquely describes the crack-tip fields as, respectively, in Eqs. 3.1 and 4.2a. An example of this is illustrated in Fig. 4.12 which shows finite element calculations on the effect of T/σ_o, varying from -1 to $+1$, on the crack-tip stress field deep within the plastic zone compared to that predicted by the HRR field (σ_o here is the flow stress) [29]. The $0 \le T/\sigma_o \le 1$ stress distributions are consistent with the HRR solution (Eq. 4.2), but negative T values ($-1 \le T/\sigma_o \le 0$) have a marked effect in lowering the crack-tip stresses.

If we examine how the T-stress varies in typical test geometries, it is found that it is negative in the middle-cracked tension geometry. This, if you recall from Section 4.3.1, is a specimen configuration which is unlikely to achieve crack-tip constraint consistent with the J-dominant HRR

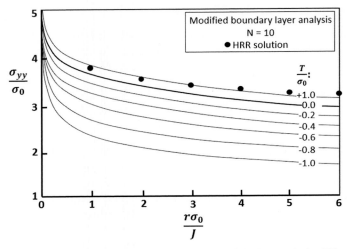

FIGURE 4.12   Finite element calculations of the effect of the T-stress $(-1 \leq T/\sigma_o \leq +1)$ on the crack-tip σ_{yy} stress as a function of normalized distance ahead of the crack tip within the plastic zone, as compared to the HRR solution. σ_o is the flow stress [29]. (reproduced with permission).

solution in the presence of plastic deformation. However, the highly constrained Prandtl-field-like configurations, such as the single-edge notched tension and bend geometries, both have positive T-stresses when deeply cracked. As shown in Fig. 4.12, these geometries with a positive T maintain a level of high crack-tip constraint with plastic deformation, which can even be retained out to fully plastic conditions. For these geometries, we can conclude that a single-parameter characterization of the crack-tip fields, such as J in terms of the HRR field, and the use of this parameter as a size- and geometry-independent descriptor of the fracture toughness, is a reasonably robust and sound approach. However, for cracked configurations where T is negative, this approach is a far more contentious.

4.7.2 Two-parameter fracture mechanics

These deliberations that the T-stress can affect the crack-tip constraint in different cracked configurations, and hence compromise the use of a single characterizing parameter, such as K_I or J, to uniquely characterize the crack-tip stress and strain fields for certain geometries, has led to the development of *two-parameter fracture mechanics* methodologies, where a second parameter, such as T, is used to characterize the crack-tip fields in addition to K_I or J [27,30]. As the effect is most pertinent with increasing degrees of plastic deformation and T is an elastic parameter, a plastic version of this concept has been developed, termed the *J-Q theory* [30], where Q characterizes the difference between the actual crack-tip stresses and those prescribed by the *J*-dominant HRR solution (where $T = 0$), i.e., it is a measure of the relative constraint at the crack tip and can be interpreted as a stress triaxiality parameter. Thus, the actual crack-tip stresses can be approximated in terms of those for the HRR solution, modified by a Q term:

$$\sigma_{ij} = (\sigma_{ij})_{HRR} + Q\sigma_o \, \delta_{ij} \, , \qquad (4.26)$$

where δ_{ij} is the Kronecker delta (=1 when $i = j$ and 0 when $i \neq j$). If Q is positive, the stress triaxiality is increased relative to the HRR ($Q = 0$) state, whereas if Q is negative, the stress triaxiality is decreased. Specifically, Q and T are uniquely related, with T and $Q = 0$ for SSY. As a more negative Q signifies a loss of stress triaxiality at the crack tip, the critical value of J_c at fracture will tend to increase. Accordingly, the approach implies that the fracture toughness needs to be specified in terms of both J_c and Q. This approach can thus facilitate the measurement of toughness in specimen geometries with low constraint, where the HRR-based description is compromised. It is therefore particularly useful for the prediction of the failure of actual structures, where cracks in ductile steels, for example, may be in complex configurations. Further details of *J-Q* two-parameter methodology can be found in the original study by O'Dowd and Shih [30].

References

[1] R.O. Ritchie, The conflicts between strength and toughness, Nat. Mater. 10 (2011) 817.
[2] J.D. Eshelby, The continuum theory of lattice defects, Solid State Phys. 3 (1956) 79.
[3] J.L. Sanders, On the Griffith-Irwin fracture theory, J. Appl. Mech. 27 (1960) 352.
[4] G.P. Cherepanov, The propagation of cracks in a continuous medium, J. Appl. Math. Mech. 31 (1967) 503.
[5] J.R. Rice, G.F. Rosengren, Plane strain deformation near a crack tip in a power-law hardening material, J. Mech. Phys. Solid. 16 (1968) 1.
[6] J.W. Hutchinson, Singular behavior at the end of a tensile crack tip in a hardening material, J. Mech. Phys. Solid. 16 (1968) 13.
[7] J.R. Rice, A path-independent integral and the approximate analysis of strain concentration by notches and cracks, J. Appl. Mech. 35 (1968) 379.
[8] A.A. Wells, Unstable crack propagation in metals: cleavage and fast fracture, Proc. Crack Propag. Symp. 1 (1961) 210. Cranfield College of Aeronautics, UK.
[9] M.L. Williams, On the stress distribution at the base of a stationary crack, J. Appl. Mech. 24 (1957) 109.
[10] J.A. Begley, J.D. Landes, The J-Integral as a fracture criterion, in: ASTM STP 514, American Society for Testing and Materials, West Conshohocken, PA, 1972, p. 1.
[11] J.D. Landes, J.A. Begley, The effect of specimen geometry on J_{Ic}, in: ASTM STP 514, American Society for Testing and Materials, West Conshohocken, PA, 1972, p. 24.
[12] F.A. McClintock, Plasticity aspects of fracture, in: H. Liebowitz (Ed.), Fracture: An Advanced Treatise, vol. 3, Academic Press, New York, NY, 1971, p. 47.

[13] J.W. Hutchinson, A Course on Nonlinear Fracture Mechanics, The Tech. Univ. of Denmark, Copenhagen, 1979.

[14] B. Gludovatz, S.E. Naleway, R.O. Ritchie, J.J. Kruzic, Size-dependent fracture toughness of bulk-metallic glasses, Acta Mater. 70 (2014) 198.

[15] ASTM Standard E1820-20, Standard Test Method for Measurement of Fracture Toughness. Annual Book of ASTM Standards, 3.01, American Society for Testing and Materials, West Conshohocken, PA, 2020.

[16] J.A. Begley, J.D. Landes, Serendipity and the J integral, Int. J. Fract. 12 (1976) 764.

[17] R.M. McMeeking, D.M. Parks, On the criteria for J-Dominance of crack tip fields in large-scale yielding, in: ASTM STP 668, American Society for Testing and Materials, West Conshohocken, PA, 1979, p. 175.

[18] T.L. Anderson, Fracture Mechanics: Fundamentals and Applications, fourth ed., CRC Press, Boca Raton, FL, 2017.

[19] A. Saxena, Nonlinear Fracture Mechanics for Engineers, CRC Press, Boca Raton, FL, 1997.

[20] V. Kumar, M.D. German, C.F. Shih, An Engineering Approach for Elastic Plastic Analysis, EPRI Rep. no. NP-1931, Electric Power Research Institute, Palo Alto, CA, 1981.

[21] J.R. Rice, P.C. Paris, J.G. Merkle, Some further results of J-Integral analysis and estimates, in: ASTM STP 536, American Society for Testing and Materials, West Conshohocken, PA, 1973, p. 231.

[22] H.A. Ernst, P.C. Paris, J.D. Landes, Estimations of J-Integral and tearing modulus T from a single specimen test record, in: ASTM STP 743, American Society for Testing and Materials, West Conshohocken, PA, 1981, p. 476.

[23] G.H. Aronson, R.O. Ritchie, Optimization of the electrical potential technique for crack growth monitoring in compact test pieces using finite element analysis, ASTM J. Test. Eval. 7 (1979) 208.

[24] P.C. Paris, H. Tada, A. Zahoor, H.A. Ernst, The theory of instability of the tearing mode of elastic-plastic crack growth, in: ASTM STP 668, American Society for Testing and Materials, West Conshohocken, PA, 1979, p. 5.

[25] P.S. Leevers, J.C. Radon, Inherent biaxiality in various fracture specimen geometries, Int. J. Fract. 19 (1982) 311.

[26] S.G. Larsson, A.J. Carlsson, Influence of non-singular stress terms and specimen geometry on small-scale yielding at crack tips in elastic-plastic materials, J. Mech. Phys. Solid. 21 (1973) 263.

[27] C. Betegon, J.W. Hancock, Two parameter characterization of elastic-plastic crack tip fields, J. Appl. Mech. 58 (1991) 104.

[28] A.H. Sherry, C.C. France, M.R. Goldthorpe, Compendium of T-stress solutions for two and three dimensional cracked geometries, Fatig. Fract. Eng. Mater. Struct. 18 (1995) 141.

[29] M.T. Kirk, R.H. Dodds, T.L. Anderson, Approximate techniques for predicting size effects on cleavage fracture toughness (J_c) using the elastic T stress, in: ASTM STP 1207, American Society for Testing and Materials, West Conshohocken, PA, 1994, p. 62.

[30] N.P. O'Dowd, C.F. Shih, Two-parameter fracture mechanics: theory and applications, in: ASTM STP 1207, American Society for Testing and Materials, West Conshohocken, PA, 1994, p. 21.

Crack-tip opening displacement (CTOD)

5.1 Introduction

Before the concept of J was introduced in the late 1970s as a fracture parameter to use for toughness measurements in the presence of some degree of plasticity, Wells in the UK proposed that the crack opening displacement, now termed the *crack-tip opening displacement* (CTOD), could be employed as a criterion for fracture in materials that were too tough to be characterized by linear-elastic fracture mechanics [1]. The belief that a critical displacement can be correlated to the onset of fracture, and hence used as a measure of toughness, is reasonable, of course. Since the crack-tip displacements, like the crack-tip stresses and strains, are unique and autonomous and characterized by K_I under linear-elastic conditions and by the HRR fields characterized by J under nonlinear-elastic conditions (provided the pertinent validity criteria are satisfied), the crack-tip displacements, respectively, for these two singular fields in Eqs. 3.1 and 4.2, are likewise so. Accordingly, the CTOD can be used as the basis of another fracture mechanics characterizing parameter. Indeed, it is used extensively in certain industries, such as the off-shore oil and gas industries, as a means of qualifying the toughness of weldments.

5.2 Calculation of the CTOD

The presence of a plastic zone ahead of a crack means that the crack tip will experience some degree of blunting, which can be described in mode I in terms of a finite opening displacement of the crack tip, δ_t, which is the CTOD. A commonly used definition is shown in Fig. 5.1 in terms of

75

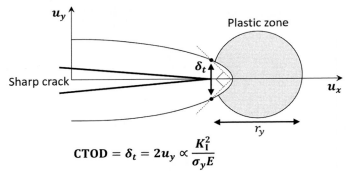

FIGURE 5.1   Definition of the crack-tip opening displacement (CTOD).

the total opening displacement measured at intersection of two orthogonal lines drawn from the crack tip to the crack flanks. This is close to the opening displacement at the location of the original crack tip prior to blunting, which has also been used as a definition.

As the CTOD is related to crack displacements that are uniquely defined by the crack-tip fields, using these singular solutions, the CTOD can be readily related to other characterizing parameters such as K_I, G, and J. Under SSY conditions (in mode I), δ_t can be determined from the displacement $2u_y$ from the linear-elastic singular solutions (Eq. 3.1b), defined at a specific distance (on the order of r_y) behind the crack tip, as per Fig. 5.1. From Eq. 3.1b, the displacement u_y can be written as:

$$u_y = \frac{4}{E'} K_I \sqrt{\frac{r_y}{2\pi}}, \tag{5.1a}$$

such that the CTOD is given in terms of the yield strength σ_y by:

$$\delta_t = 2u_y = \frac{4}{\pi} \left(\frac{K_I^2}{\sigma_y E} \right) = \frac{4}{\pi} \left(\frac{G}{\sigma_y} \right), \tag{5.1b}$$

where the plastic-zone radius r_y is given by Eq. 3.4, and $E' = E$ (Young's modulus) in plane stress and $E/(1 - \nu^2)$ in plane strain (ν is Poisson's ratio).

A more commonly used relationship to describe the CTOD is based on the so-called Dugdale/Barenblatt strip-yield model [2,3], which models the plastic zone as a semi-cohesive (strip yield) zone ahead of the crack tip. Using this model, the CTOD is given by the displacement at the end of the strip yield zone as:

$$\delta_t \approx \left(\frac{K_I^2}{\sigma_y E} \right) \approx \left(\frac{G}{\sigma_y} \right), \quad \text{(plane stress)} \tag{5.2a}$$

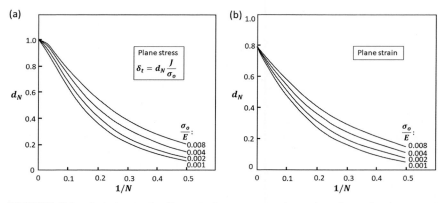

FIGURE 5.2 Variation in the dimensionless constant, d_N, in the relationship between J and the crack-tip opening displacement (CTOD) on the nonlinear-elastic HRR field in (a) plane stress and (b) plane strain [4] (reproduced with permission).

$$\delta_t \approx \frac{1}{2}\left(\frac{K_I^2}{\sigma_y E'}\right) \approx \frac{1}{2}\left(\frac{G}{\sigma_y}\right). \quad \text{(plane strain)} \quad (5.2b)$$

However, the most detailed analysis of the relationship of the CTOD to the other characterizing parameter is the nonlinear-elastic analysis of Shih [4] on how δ_t is related to the J-integral, again using twice the u_y crack displacements but now defined by the HRR field (Eq. 4.2c). Shih's analysis showed that:

$$\delta_t \approx \frac{d_N J}{\sigma_o}, \quad (5.3)$$

where σ_o is the flow stress and d_N is a dimensionless constant that varies between 0.1 and 1.0 for plane stress and 0.1 and 0.8 for plane strain, depending upon the strain hardening coefficient N and more weakly on the yield strain σ_o/E. The variation in d_N with these factors, adapted from Shih's paper [4], is shown in Fig. 5.2. The fact that δ_t can be uniquely related to J in Eq. (5.3) based on the HRR singular fields provides the basis for regarding the CTOD, like J, as a valid characterizing parameter for fracture under nonlinear-elastic conditions.

5.3 Measurement of the CTOD

Beginning with the British Standard BS5762 that first appeared in 1979, there have been several standards published for the measurement of the CTOD, although the one in current use is the ISO Standard 12135 [5] (there is also a companion ISO Standard 15653 specifically for weldments).

The general method of estimating the CTOD at the tip of an initially atomically sharp crack is to use the so-called "hinge method" by measuring the crack mouth opening displacement (CMOD), V_p, and then deducing the corresponding displacement at the crack tip on the assumption that the sample rotates as two rigid halves about a hinge point located at some fraction of the remaining uncracked ligament [5]. The set-up for a three-point bend specimen, containing a crack of length a and a remaining uncracked ligament of $b = (W - a)$, is shown in Fig. 5.3. For such an SE(B) sample, V_p is measured using a displacement or clip gauge mounted on knife-edges (of height z) attached to the top surface of the specimen. Then if the hinge point is located at a distance rb below the crack tip, where r is a dimensionless constant between 0 and 1, then the displacement at the crack tip, δ_t, can be estimated using similar triangles as:

$$\frac{\delta_t}{rb} = \frac{V_p}{rb + a + z} \ . \tag{5.4}$$

In the ISO 12135 Standard [5], the crack-tip opening displacement δ_t is separated into elastic and plastic parts $(\delta_t = \delta_{el} + \delta_{pl})$, where δ_{el} is computed from Eq. 5.2 using the linear-elastic stress intensity K_I and δ_{pl} from Eq. (5.4), i.e., for plane strain:

$$\delta_t = \frac{K_I^2}{2 \, \sigma_y \, E'} + \frac{r_p b_o V_p}{r_p b_o + a_o + z} \ , \tag{5.5}$$

where σ_y is the yield stress, E' is the plane strain Young's modulus $E' = E/(1 - \nu^2)$, a_o and b_o are the respective initial values of the crack length and uncracked ligament, and r_p is the plastic rotational constant, which is given as $r_p = 0.4$ for the SE(B) sample and 0.46 for the C(T) sample (although for Eq. (5.5) to be representative for this geometry, V_p needs to be measured at the load-line, i.e., where $V_p = \Delta_{LL}$).

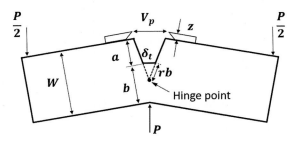

FIGURE 5.3   Measurement of the crack-tip opening displacement δ_t from the crack mouth opening displacement V_p, using the "hinge method," for a pre-cracked (three-point) single-edged notched bend specimen.

To determine the $\delta_t(\Delta a)$ R-curve, as with the $J_R(\Delta a)$ curve in Section 4.6.2, the CTOD needs to be calculated at each measured increment of crack extension, such that for the ith measurement point, the CTOD is calculated in terms of the current values of the stress intensity $K_{I[i]}$ and crack length $a_{[i]}$:

$$\delta_{t[i]} = \frac{K_{I[i]}^2}{2\,\sigma_y\,E'} + \frac{(1 - r_p)\Delta a_{[i]} + r_p\,b_o V_{p[i]}}{r_p b_{[i]} + a_{[i]} + z} \, , \qquad (5.6)$$

where $\Delta a_{[i]} = a_{[i]} - a_o$. In terms of the ISO 12135 Standard [5], the $\delta_t(\Delta a)$ R-curve is considered to be valid provided two criteria are satisfied, specifically that:

$$\Delta a_{\max} \leq 0.25 b_o \text{ and } \delta_{t,\max} \leq \min(B\,/\,30; a_o\,/\,30; b_o\,/\,30) \, , \qquad (5.7)$$

where B is the sample thickness.

The final requirement is to assess at what point of the load-displacement curve will the CTOD be measured. Several cases are described in the Standard [5], namely:

- if no stable cracking precedes final failure, δ_t is defined at fracture instability as δ_c.
- if stable cracking does precede final failure, δ_t is again defined at fracture instability as δ_u.
- if no instability occurs prior to maximum load, δ_t can be defined at maximum load as δ_m.
- δ_t is best defined at crack initiation, as δ_i, but this must be measured from the R-curve.

5.4 Crack-tip opening angle (CTOA)

Unlike fracture parameters such as the stress intensity and the J-integral, the CTOD is a physically realistic entity. One could envision how the crack-tip displacement could relate to the strain between two particles to induce ductile fracture via microvoid coalescence; indeed, there are models, e.g., ref [6], that propose for this scenario that the critical CTOD at fracture is given by $\delta_{Ic} \approx (0.5-2.0)d_p$, where d_p is the mean spacing of the particles.

Moreover, the slope of the $\delta_t(\Delta a)$ R-curve, $d\delta_t/da$, is the *crack-tip opening angle*, ϕ_t, which can be utilized as another measure of the crack-growth fracture toughness. This parameter has only been used sparingly for characterizing macro-scale fracture properties, although there has been a recent work on its use for crack initiation and arrest in pipeline steels [7]. One problem is that it can be difficult to define numerically depending upon the form of the crack-tip opening profile. However, like the CTOD, it is a physically realistic entity which can be related to the physics of stable crack growth. Indeed, a micro-mechanical model [8] for the stable growth of a mode I tensile crack based on a constant CTOA criterion is described in Section 6.3.3.

References

[1] A.A. Wells, Unstable crack propagation in metals: cleavage and fast fracture, Proc. Crack Propag. Symp 1 (1961) 210. Cranfield College of Aeronautics, UK.
[2] D.S. Dugdale, Yielding in steel sheets containing slits, J. Mech. Phys. Solid. 8 (1960) 100.
[3] G.I. Barenblatt, The mathematical theory of equilibrium cracks in brittle fracture, Adv. Appl. Mech. VII (1962) 55.
[4] C.F. Shih, Relationship between the J-integral and the crack opening displacement for stationary and extending cracks, J. Mech. Phys. Solid. 29 (1981) 305.
[5] International Standard ISO 12135, Metallic Materials — Unified Method of Test for the Determination of Quasistatic Fracture Toughness, International Standards Organization, Geneva, 2016 (2016).
[6] J.R. Rice, M.A. Johnson, The role of large crack tip geometry changes in plane strain fracture, in: M.F. Kanninen, W.F. Adler, A.R. Rosenfield, R.I. Jaffee (Eds.), Inelastic Behavior of Solids, McGraw-Hill, New York, NY, 1970, p. 641.
[7] J.Q. Wang, J. Shuai, Measurement and analysis of crack tip opening angle in pipeline steels, Eng. Fract. Mech. 79 (2012) 36.
[8] J.R. Rice, W.J. Drugan, T.-L. Sham, Elastic-plastic analysis of growing cracks, in: ASTM STP 700, American Society for Testing and Materials, West Conshohocken, PA, 1980, p. 189.

6

Micromechanics modeling of fracture

6.1 Introduction

The power of continuum mechanics, which underlies the fundamentals of fracture mechanics, is that it can be readily applied to brittle as well as ductile fractures, not to mention environmentally-assisted cracking, creep and fatigue failures, etc., as the fracture criteria, e.g., $K_I = K_{Ic}$, $G = G_{Ic}$, or $J = J_{Ic}$, are *global* criteria and hence independent of the micromechanisms of failure. Nevertheless, in many cases, *local* models based on microstructural features have been developed for fracture modes, such as critical stress models for cleavage fracture [1] and critical strain models for microvoid coalescence [2,3]. This is arguably the essence of the understanding of fracture if such microscale mechanistic models can be embedded into fracture mechanics-based global criteria. This approach is known as the micromechanical modeling of fracture. Whereas with continuum models, one really only needs to know that the characterizing parameter, e.g., K_I or J, actually characterizes the crack-tip stresses and strains; for micromechanical modeling one needs to know the actual distribution of stresses and strains at the crack tip that are pertinent to the specific mechanism of fracture.

6.2 Fracture mechanisms

Before we describe the micromechanistic models based on these crack-tip fields, we will briefly explain the primary mechanisms of ductile and brittle (overload) fracture.

Introduction to Fracture Mechanics
https://doi.org/10.1016/B978-0-323-89822-5.00009-8

81

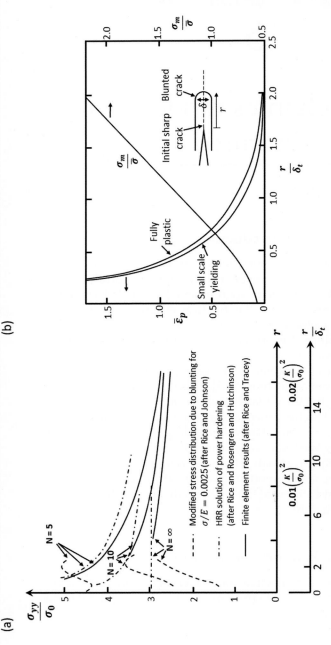

FIGURE 6.1 Crack-tip fields in plane strain showing (a) the distribution of the local tensile σ_{yy} stress, and (b) local equivalent plastic strain $\bar{\epsilon}_p$, at distance $r = x$ ($\theta = 0$) directly ahead of the crack tip. The stress distribution in (a) is the HRR asymptotic field for a power-hardening solid [4,5] (Eq. 4.2a), with a finite element solution [6] shown for comparison, both truncated as $r \rightarrow 0$ by Rice and Johnson's near-tip blunting solution [7] (for an initial yield strain σ_0/E of 0.0025). The abscissa is normalized with respect to both $(K/\sigma_0)^2$ and δ_t, the CTOD; the ordinate is normalized with respect to the flow stress σ_0. The strain distribution in (b) is based on the finite-geometry blunting solutions for both small-scale yielding and fully plastic conditions [7,8]. Also shown is the corresponding variation in the stress-state, defined in terms of the ratio of the hydrostatic stress and the equivalent stress, $\sigma_m/\bar{\sigma}$.[1] The abscissa is normalized with respect to δ_t. *Reproduced with permission.*

[1]The hydrostatic stress is the mean normal stress: $\sigma_m = \frac{1}{3}(\sigma_{xx} + \sigma_{yy} + \sigma_{zz})$. The equivalent (Mises) stress is given by

$$\bar{\sigma} = \left\{ \frac{1}{2}\left[(\sigma_{xx} - \sigma_{yy})^2 + (\sigma_{yy} - \sigma_{zz})^2 + (\sigma_{zz} - \sigma_{xx})^2 \right] + 3\left(\sigma_{xy}^2 + \sigma_{yz}^2 + \sigma_{zx}^2 \right) \right\}^{\frac{1}{2}}, \text{ and the equivalent plastic strain is given by}$$

$$(d\epsilon_p)^2 = \frac{4}{9}\left\{ \frac{1}{2}\left[(d\epsilon_{xx} - d\epsilon_{yy})^2 + (d\epsilon_{yy} - d\epsilon_{zz})^2 + (d\epsilon_{zz} - d\epsilon_{xx})^2 \right] + 3\left(d\epsilon_{xy}^2 + d\epsilon_{yz}^2 + d\epsilon_{zx}^2 \right) \right\}.$$

The first question to ask is what is the difference between ductile and brittle fracture? This is principally associated with the extent of plastic, or more generally inelastic, deformation, involved in the fracture mechanism. Indeed, inelasticity is key to promoting intrinsic toughness in materials, as it acts to dissipate locally high stresses which might otherwise lead to fracture.[2] In fundamental terms, brittle fracture involves the breaking of an atomic bond, e.g., in the high stress region at a crack tip, with no plasticity involved. With ductile fracture, conversely, a dislocation is first emitted from the tip (e.g., in metallic structures), which leads to the blunting of the crack. However, the actual modes of fracture in real materials are invariably a compromise between these two extremes, i.e., there is generally always some inelasticity involved even for brittle fracture — it is just a question of the degree. These modes are shown in Fig. 6.2 as electron micrographs of the classical fracture mechanisms, i.e., ductile fracture by microvoid coalescence and brittle fracture by quasi-cleavage, intergranular cracking, and transgranular cleavage.

6.2.1 Ductile fracture

The typical mode of ductile fracture in structural materials involves the coalescence of microvoids formed at second-phase particles such as inclusions or precipitates, shown by the dimples on the fracture surface in Fig. 6.2a. This fracture process is invariably *strain-controlled* where the voids are initially generated by particles debonding from the matrix, although they can also be created by particle fracture; these voids then grow under the imposed strain, aided by triaxial stresses, e.g., at a crack tip, until they coalesce either by impingement or more likely by plastic

---

[2] Plastic deformation can take many forms in different material systems. The most well-known mechanism is dislocation motion in crystalline metallic structures, but deformation can also occur in these materials by twinning at low temperatures and diffusional flow at high temperatures. Dislocation motion in ceramics is more difficult due to their strong directional bonding, but ceramics can deform plastically at high homologous temperatures by dislocation motion and by diffusional flow. In rocks and some ceramics, inelastic deformation can be associated with microcracking, whereas in polymers deformation is invariably due to the stretching and rotation of molecular bonds. In biological materials, such as skin and bone, inelasticity takes place by the stretching and sliding between collagen fibrils, in sea shells by the micron-scale motion between the mineral bricks from which they are constructed. All these mechanisms consume energy and serve to dissipate locally high stresses, thereby enhancing resistance to fracture.

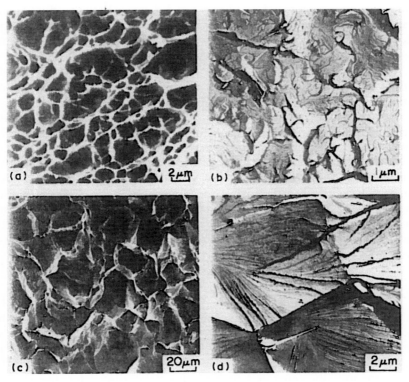

FIGURE 6.2 Classical fracture morphologies showing (a) microvoid coalescence, (b) quasi-cleavage, (c) intergranular cracking, and (d) transgranular cleavage. Fractographs (a) and (c) were obtained using scanning electron microscopy whereas (b) and (d) are from transmission electron microscopy replicas [9]. *Reproduced with permission.*

instability, e.g., necking, in the ligaments between adjacent voids. As extensive plasticity is involved, microvoid coalescence is generally concomitant with high toughness fractures.

6.2.2 Brittle fracture

There are several forms of brittle fracture, as shown in Fig. 6.2b–d, most notably transgranular cleavage (Fig. 6.2d) where the catastrophic fracture ensues along low energy, crystallographic "cleavage planes", such as the {100} planes in steels or {111} planes in silicon and diamond. This process is distinctly locally *stress-controlled*, and in steels, where much research has been conducted, generally involves the pile-up of dislocations against a carbide particle located at a grain boundary which fractures and transmits a crack into the ferrite matrix where it propagates catastrophically. Cleavage fractures tend to be "shiny" in appearance[3] and to display "river markings" on the cleavage facets which radiate out from the point of origin of the fracture in each grain; these result from the local "cliffs" between adjacent cracks that form on slightly different planes.

There is a somewhat similar process, termed quasi-cleavage (Fig. 6.2b), which is often seen in martensitic steels where the fracture originates from a cracked particle in the middle of a grain (i.e., the river markings radiate out from the center rather than from the edge of the facet). Quite frankly, quasi-cleavage is used as a "catch-all" term to describe nominally brittle fractures that do not look exactly like cleavage, but the term strictly refers to a transitional mode of fracture where cleavage facets form and then link together by ductile fracture.

The other major mode of brittle fracture is intergranular fracture (Fig. 6.2c), which is a dangerous mechanism of failure invariably resulting from segregated species to the grain boundaries. This can be associated with decohesion due to impurities such as S and P in steels or hydrogen in metals, where it can be considered as a locally stress-controlled fracture. It is also a common mode of failure for most creep fractures where the intergranular facets tend to be "pock-marked" due to cavitation around particles in the boundaries.

6.3 Cleavage and ductile fracture models

6.3.1 RKR model for cleavage fracture

In view of the specificity of micromechanical models to particular fracture mechanisms for particular microstructures, a complete microscopic/macroscopic characterization of toughness with such an approach has only been achieved in a few simplified cases. One of the first of these models was the Ritchie-Knott-Rice (RKR) model [1] for transgranular cleavage which was originally derived to predict the plane-strain, crack-initiation toughness, K_{Ic}, by slip-initiated cleavage in ferritic steels. These authors embedded a local stress-controlled criterion within a crack-tip stress distribution in Fig. 6.1a defined by the HRR field (Eq. 4.2a), and proposed that the onset of fracture could be modeled in terms of a local critical fracture stress, σ_f^*, being exceeded over a characteristic

---

[3] In past times, when people analyzed catastrophic cleavage fractures and saw the shiny facets on the fracture surfaces, they would conclude that the material fractured " .... because it had crystallized," which of course is an absurdly erroneous assessment.

6. Micromechanics modeling of fracture

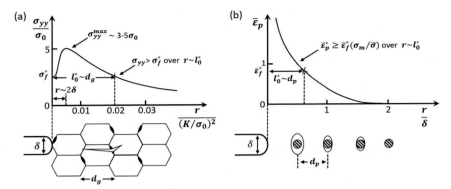

FIGURE 6.3 Schematic idealization of microscopic fracture criteria pertaining to (a) RKR critical stress-controlled model for cleavage fracture and (b) stress-state modified critical strain-controlled model for microvoid coalescence [9]. *Reproduced with permission.*

microstructural distance, l_o^*, ahead of the crack tip, which they found to be approximately two grain diameters $(l_o^* \approx 2d_g)$ (Fig. 6.3a). Their derivation is outlined below [1,9].

For a Ramberg-Osgood solid, with a constitutive law given in Eq. 4.1 by $\bar{\varepsilon}/\varepsilon_0 = \alpha\,(\bar{\sigma}/\sigma_0)^N$, the HRR crack-tip stress distribution (Eq. 4.2a) is given by Ref. [4,5]:

$$\sigma_{ij} = \sigma_o \left[\frac{J}{\alpha\sigma_o\varepsilon_o I_N r} \right]^{1/N+1} \tilde{\sigma}_{ij}(\theta,\, N)\ . \tag{6.1}$$

Using Eq. (6.1) to derive the local tensile stress σ_{yy} directly ahead of the crack tip (at $\theta = 0$) at fracture, when the global fracture criterion is given by $K_I = K_{Ic}$, i.e., when $J = J_{Ic} = K_{Ic}^2/E'$, gives:

$$\sigma_{yy} = \sigma_o \left[\frac{K_{Ic}^2}{\alpha I_N \sigma_o^2 r} \right]^{1/N+1} \tilde{\sigma}_{ij}(\theta,\, N)\ , \tag{6.2}$$

where the flow stress $\sigma_o = E\varepsilon_o$.

Applying the RKR criterion (Fig. 6.3a), where the *local* failure at the onset of crack initiation is when the fracture stress σ_f^* is exceeded over the characteristic distance l_o^*:

$$\sigma_{yy} > \sigma_f^* \quad \text{over } r = l_o^*, \text{ where } \quad l_o^* \sim d_g\ , \tag{6.3}$$

and then substituting into Eq. (6.2) gives an expression for the fracture toughness K_{Ic} for transgranular cleavage fracture:

$$K_{Ic} = \left[\frac{(\alpha I_N)^{\frac{1}{2}}}{\tilde{\sigma}_{yy}(\theta,N)^{\frac{N+1}{2}}} \right] \left(\frac{\sigma_f^{*\frac{N+1}{2}}}{\sigma_o^{\frac{N-1}{2}}} \right) l_o^{*\frac{1}{2}}\ , \tag{6.4}$$

where the first term in square brackets simply pertains to dimensionless factors in the HRR solution, namely the angular function $\tilde{\sigma}_{yy}(\theta, N)$ and the integration constant I_N, which are readily available in the tabulations [5,10], with α of order unity in the constitutive law. This expression indicates that the cleavage fracture toughness is a direct function of the local fracture stress σ_f^*, which is of course expected, but it also indicates that the toughness is inversely proportional to the strength of the material and will be lower in high work-hardening materials, where the stresses become higher at the crack tip. The inverse relationship between strength and toughness is widely observed in many materials [11]. This largely results because higher toughness can be promoted by an enhanced role of plastic deformation, which naturally is more prevalent in lower strength materials. K_{Ic} will also be directly proportional to the square root of the characteristic distance, i.e., to the square root of the grain size d_g, but sometimes this is obscured by the fact that the grain size also affects the strength and ductility.

In mild steels, with ferrite/carbide microstructures, the characteristic distance was found to be on the order of the spacing of the void initiating grain-boundary carbides, i.e., typically two grain diameters [1,12], although different size scales have been found when the analysis is applied to other materials [13]. The model has been found to be particularly successful both in quantitatively predicting cleavage fracture toughness values in a wide range of microstructures and furthermore in rationalizing the influence on K_{Ic} of such variables as temperature [1,14,15], strain rate [14−16], and neutron irradiation [14,16]. Indeed, adaptations of the RKR model have been suggested for other fracture modes, including inter-granular cracking and hydrogen-assisted fracture [17].

The RKR model has also been reconstituted on a stochastic basis using weakest-link statistics [12,18], which can provide important information on the likely scatter in toughness values, and further can offer some insight into the meaning of the characteristic distance. In simple terms, this distance represents the location from the crack tip where there is the maximum probability of fracture initiating. As the σ_{yy} stress peaks very close (within two CTODs) to the crack tip and then decays with increasing distance r away from the tip, one might expect that the critical fracture event would occur very close to the tip. But there is also the statistical probability of finding the "weakest link," e.g., the most brittle particle, and this increases with increasing distance from the crack tip, i.e., with an increasing sampling volume. So the highest stresses are close to the tip, whereas the highest probability of finding a weakest link is farthest away from the tip, such that the maximum probability for the initial (triggering) fracture event is in-between − a compromise between high stresses and finding the weakest link − which turns out in low strength ferritic steels to be on the order of a couple of grain sizes ahead of the crack tip.

6.3.2 Critical strain model for ductile fracture

For the initiation of ductile fracture by microvoid coalescence (Fig. 6.2a), a simple criterion [7] is that the critical CTOD, δ_{Ic}, must exceed half the mean void-initiating particle spacing, i.e., $\delta_{Ic} \approx (0.5-2.0)d_p$, where d_p is the average particle spacing. This model is based on the notion that, in nonhardening materials, this would take place when the void sites are first enveloped by the intense strain region at distance $r \sim 2\delta_t$ from the crack tip. However, an improved model involves a strain-based version of RKR, first proposed by McClintock [2], involving a critical strain across a characteristic distance ahead of the crack tip [3], which is related to d_p (Fig. 6.3b), as described below.

For Ramberg-Osgood material, the HRR crack-tip strain distribution (Eq. 4.2b) is given by:

$$\varepsilon_{ij} = \alpha\varepsilon_o \left[\frac{J}{\alpha\sigma_o\varepsilon_o I_N r}\right]^{N/N+1} \tilde{\varepsilon}_{ij}(\theta, N) . \qquad (6.5)$$

For large N, $N/(N+1) \to 1$, such as the local tensile strain distribution, directly ahead of the crack tip (at $\theta = 0$), can be approximated as:

$$\bar{\varepsilon}_p = \left[\frac{\tilde{\varepsilon}_{ij}(\theta, N)}{\alpha I_N}\right] \frac{J}{\sigma_o r} . \qquad (6.6)$$

At fracture, $J = J_{Ic}$ is the global criterion, whereas the local criterion is that the local strain, given by Eq. (6.6), exceeds a critical fracture strain (ductility) over a characteristic distance of l_o^* which is now related to the particle spacing, d_p (Fig. 6.3b), i.e.,

$$\bar{\varepsilon}_p > \varepsilon_f^* \text{ over } r = l_o^*, \text{ where } l_o^* \sim d_p ; \qquad (6.7)$$

substituting this into Eq. (6.6) gives an expression for the fracture toughness J_{Ic} for ductile fracture:

$$J_{Ic} = \left[\frac{\alpha I_N}{\tilde{\varepsilon}_{ij}(\theta, N)}\right] \sigma_o \bar{\varepsilon}_f^* l_o^* , \qquad (6.8a)$$

or in terms of K_{Ic}:

$$K_{Ic} = \left[\frac{\alpha I_N}{\tilde{\varepsilon}_{ij}(\theta, N)}\right]^{\frac{1}{2}} \left(E' \sigma_o \bar{\varepsilon}_f^* l_o^*\right)^{\frac{1}{2}} , \qquad (6.8b)$$

where the first term in square brackets again pertains to a dimensionless function of α, the angular function and integration constant in the HRR solution (and E' is Young's modulus).

This simple relationship implies that the plane-strain J_{Ic} fracture toughness for ductile fracture is a simple function of the product of the strength, ductility, and a characteristic dimension related to the particle spacing.

There is, however, one subtlety that should be taken care of. The ductility $\bar{\varepsilon}_f^*$ in these relationships is not the ductility measured in a uni-axial tensile test but rather the ductility pertaining to the highly triaxial conditions ahead of the crack tip. This triaxiality (or stress-state) is measured in terms of the ratio of hydrostatic to equivalent stress, $\sigma_m / \bar{\sigma}$, which from Fig. 6.1b can be seen to increase throughout the crack-tip strain field. As void growth can occur far more readily in a triaxial field, the plane-strain ductility (where $\sigma_m/\bar{\sigma} = 2$) can be as much as an order of magnitude smaller than the ductility measured in uniaxial tension (where $\sigma_m/\bar{\sigma} = 1/3$). This is predicted by the Rice and Tracey analysis [19] for void growth ahead of a crack tip in a nonhardening solid, where, in terms of the particle spacing d_p and size D_p, the ductility $\bar{\varepsilon}_f^*$ can be seen to be an inverse exponential function of the stress-state:[4]

$$\bar{\varepsilon}_f^* \approx \frac{\ln(d_p/D_p)}{0.28 \exp(1.5\,\sigma_m/\bar{\sigma})} \;. \tag{6.9}$$

Accordingly, to use the strain-based toughness model correctly, one needs to measure the ductility as a function of $\sigma_m/\bar{\sigma}$, which can be per-formed using circumferentially-notched tensile specimens (with varying notch size and root radius), and to then match the value of $\bar{\varepsilon}_f^*$ to the particular stress-state at the location of the initial fracture, i.e., at $r = l_o^*$ [14].

So how well do these models work? Shown in Fig. 6.4 is the plane-strain fracture toughness K_{Ic} of SA533B nuclear pressure vessel steel as a function of temperature, showing the brittle-to-ductile transition at a temperature above $-30°C$. On the lower shelf where the steel fails by cleavage, the RKR criterion (Eq. 6.4) can predict the toughness with a characteristic distance of about 2 to 4 grain diameters (essentially the bainite packet size), whereas on the upper shelf where the steel fails by ductile fracture, the critical strain model (Eq. 6.8) can predict the tough-ness with a characteristic distance of roughly 5 to 6 times the inclusion spacing [14]. (The alloy contains ~ 0.12 vol.% of manganese sulfide and aluminum oxide inclusions, roughly $5-10$ μm in diameter).

6.3.3 Critical CTOA model for ductile crack growth

We have shown in Section 6.3.2 that the concept of a critical strain being attained over some characteristic dimension directly ahead of a stationary crack can be used to predict the mode I crack-initiation toughness, but this

---

[4] It should be noted that the Rice and Tracey analysis [19] tends to overestimate the ductility as it assumes that fracture occurs by the simple impingement of the growing voids and thus ignores any prior coalescence due to shear banding by strain localization. However, the analysis correctly displays the dependence of the ductility on stress-state ($\sigma_m/\bar{\sigma}$) and purity (d_p/D_p).

6. Micromechanics modeling of fracture

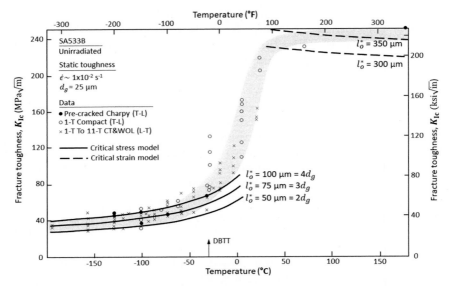

FIGURE 6.4 Comparison of the experimentally measured fracture toughness K_{Ic} for SA533B nuclear pressure vessel steel with predicted values based on the RKR critical stress model for cleavage fracture (Eq. 6.4) on the lower shelf at temperatures less than the ductile/brittle transition temperature (DBTT), and on the stress-state modified critical strain model for microvoid coalescence (Eq. 6.8) on the upper shelf above the DBTT [14]. *Reproduced with permission.*

approach is not as amenable for a growing mode I crack as the regions of intense strain are directly above and below the crack plane. However, alternative local failure criteria have been proposed for initiation and continued growth of plane-strain tensile cracks, with a geometrically similar crack profile very near the crack tip. Of particular interest is a model [20] formulated on the basis of a constant crack-tip opening angle (CTOA) ϕ_t for crack growth. As noted in Section 5.4, the CTOA represents the slope of the CTOD R-curve ($\phi_t = d\delta_t/da$), and is thus a viable measure of the crack-growth toughness. The model is shown schematically in Fig. 6.5. The CTOD at the advancing crack tip δ_p remains constant, whereas the CTOD, δ, at the original crack tip is increased by the amount of opening (δ_p) to advance the ductile crack one particle spacing $l_o^* \sim d_p$ for each increment of crack growth. Since the CTOA, $\phi = \arctan(\delta_p/2d_p)$, this corresponds to crack growth governed by a critical δ_p at a characteristic distance $l_o^* \sim d_p$ behind the crack tip, i.e., *to a constant CTOA.* Although a constant CTOA is a feasible local criterion for stable crack advance in a rigid-plastic solid, elastic-plastic analyses can result in crack-face profiles with a vertical tangent immediately at the crack tip,

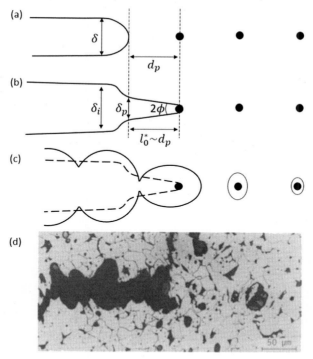

FIGURE 6.5   Model for stable growth of a mode I crack by microvoid coalescence [20], showing (a) blunted crack tip, (b) crack growth to the next inclusion based on constant CTOA (ϕ) or on critical CTOD (δ_p) at a distance $\left(l_o^* \sim d_p\right)$ behind the crack tip (d_p is the inclusion spacing), (c) morphology of resulting fracture surface [9], and (d) fractographic section [21] through a ductile crack growing via the coalescence of voids in mild steel. *Figures reproduced with permission.*

which makes the CTOA impossible to define numerically. Nevertheless, the CTOA is a useful local parameter to characterize the crack-growth toughness in ductile materials that remains relatively unexplored.

6.4 Intrinsic vs. extrinsic toughening

An Ashby plot of the toughness vs. strength of numerous classes of materials is shown in Fig. 6.6a. The mechanisms that we have examined so far to develop such toughness have been primarily for ductile materials where plasticity is the prime contribution. But what about brittle materials, such as ceramics, which invariably cannot be toughened by this means? To provide a context for this, it is useful to consider the process of fracture as a "conflict" — in fracture mechanics terms, as a mutual

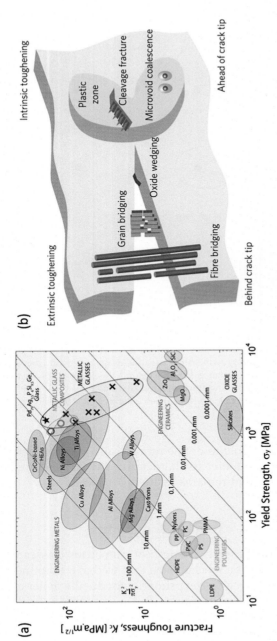

FIGURE 6.6 Conflicts of strength vs. toughness, showing (a) an Ashby plot of the strength–toughness relationships for engineering materials. Diagonal lines show the plastic-zone size, $K_c^2/\pi\sigma_y^2$, where K_c is the fracture toughness and σ_y the yield strength. The filled star and purple open circles refer, respectively, to a Pd-glass and several metallic-glass composites, which are among the most damage-tolerant materials, as compared with monolithic glasses (the *crosses*). Also shown in the orange region are CrCoNi-based high-entropy alloys, which can be even tougher. (b) Schematic illustration showing how strength and fracture behavior can be considered in terms of intrinsic (plasticity) versus extrinsic (shielding) toughening mechanisms associated with crack extension. Illustration shows mutual competition between intrinsic damage mechanisms, which act ahead of the crack tip to promote crack advance and extrinsic crack-tip-shielding mechanisms, which act at, or primarily behind the tip, to impede crack advance. Intrinsic toughening results essentially from plasticity and enhances a material's inherent damage resistance; as such, it increases both the crack-initiation and crack-growth toughnesses. Extrinsic toughening acts to lower the local stress and strain fields at the crack tip; as it depends on the presence of a crack, it affects only the crack-growth toughness, specifically through the generation of a rising R-curve [11]. *Figures reproduced with permission.*

competition between what can be termed intrinsic damage processes that operate ahead of the tip of a crack to promote its propagation, and extrinsic crack-tip-shielding mechanisms that act mostly behind the crack tip to inhibit this [11,22] (Fig. 6.6b). Damage mechanisms depend on the nature of the nano/microstructure and involve such processes as the cracking or debonding of a second phase within the process zone ahead of a crack tip. *Intrinsic toughening* is therefore associated with making these processes more difficult and, as noted above, is primarily related to plasticity, that is, enlarging the plastic zone; as such it is effective against both the initiation and propagation of cracks. With *extrinsic toughening*, conversely, the inherent fracture resistance of the material is unchanged; instead, mechanisms such as crack bridging and crack deflection act at, or more importantly in the wake of the crack to reduce (shield) the local stresses and strains actually experienced at the crack tip − stresses and strains that would otherwise have been used to extend the crack. Indeed, considering a broad class of materials, extrinsic mechanisms can be quite diverse, involving such processes as crack bridging by unbroken fibers or a ductile phase in composites, the frictional interlocking of grains during intergranular fracture in monolithic ceramics, and the presence of collagen fibrils spanning microcracks in teeth and bone. By operating principally in the crack wake, extrinsic mechanisms (unlike intrinsic mechanisms) are only effective in resisting crack growth, i.e., they only can affect the crack-growth toughness − as there has to be a crack for them to operate, they can have no effect on crack initiation. Moreover, their effect is dependent on crack length (or size). Accordingly, a major consequence of this is rising R-curve toughness behavior.

6.4.1 Toughening in metallic materials

As noted above, metals develop their toughness primarily from intrinsic mechanisms, with crack-tip plasticity as the dominating factor. With fracture, plastic deformation acts to reduce the stress intensification at the crack tip by inducing crack-tip blunting through the emission of dislocations (ductile behavior), as opposed to local decohesion by simply breaking an atomic bond there (brittle behavior). The subsequent interaction of dislocations with inhomogeneities in the microstructure then provides the mechanisms of damage, in the form of pile-ups at grain boundaries, or at second-phase particles causing them to crack or decohere from the matrix. As described in the micromechanistic models in Section 6.3, the resulting fracture can be associated with only limited plastic deformation and be (nominally) brittle, i.e., locally stress-controlled cleavage or intergranular fracture, which invariably lead to low toughness, or involve more extensive plastic deformation to cause ductile fracture (microvoid coalescence), which is locally strain-controlled and in metals invariably results in much higher toughness.

As plasticity is the prime contributor to the intrinsic toughening of metallic materials, they are generally subject to the conflict between strength and toughness — the stronger the material, the less plasticity is available for toughening. For this reason, most safety-critical structural applications, from steel pressure vessels and pipelines to aluminum alloy airframes, are manufactured from the comparatively lower-strength versions of these alloys to avoid problems from premature failure.

There are instances, though, where the conflict of attaining strength and toughness can be overcome. Although lower strain hardening in metallic structures is generally preferential for the brittle fracture toughness as it limits the elevation of stresses at a crack tip, a steady source of high strain hardening is particularly effective for the ductile fracture toughness, as this clearly enhances the strength yet at the same time delays plastic instability (necking) which serves to increase the uniform ductility. Notable examples of this are twinning-induced plasticity (TWIP) steels and CrCoNi-based high-entropy alloys, where the predominance of deformation twinning in inducing plastic deformation creates significant strain hardening; indeed, high-entropy alloys display some of the high toughnesses on record, without compromise to strength [23], as shown in Fig. 6.6a.

6.4.2 Toughening in ceramic materials

In contrast to metallic materials, at low homologous temperatures most structural ceramics, such as Al_2O_3, ZrO_2, SiC, and Si_3N_4, suffer from almost a complete absence of plastic deformation; this is due to the lack of mobile dislocation activity (due to their high Peierls force from strong directional bonding), although other modes of inelastic deformation, such as microcracking and in situ phase transformation, can provide limited alternative deformation mechanisms. The implications from this are that ceramics are inherently brittle with an extreme sensitivity to flaws. Actually, they are essentially impossible to toughen intrinsically; in the absence of extrinsic shielding mechanisms, fracture invariably occurs catastrophically (with crack initiation concomitant with instability) by cohesive bond breaking at the crack tip with a resulting very low (intrinsic) toughness of roughly 1–3 MPa√m.

Toughening ceramics, as with virtually all brittle materials, must be achieved extrinsically, i.e., through the use of microstructures which can promote crack-tip shielding mechanisms such as crack deflection, in situ phase transformations, constrained microcracking (although this mechanism is generally not too potent), and most importantly crack bridging [24] (Fig. 6.7). Extrinsic mechanisms inherently result in R-curve behavior

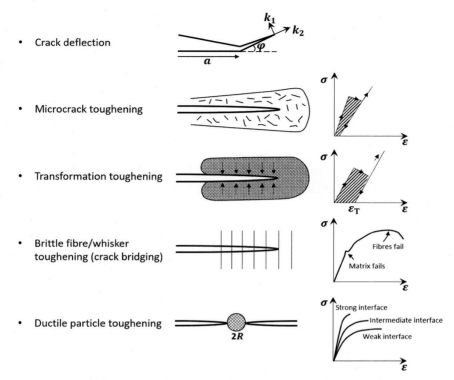

- Crack deflection

- Microcrack toughening

- Transformation toughening

- Brittle fibre/whisker toughening (crack bridging)

- Ductile particle toughening

FIGURE 6.7   Schematic illustration; several of the primary toughening mechanisms in ceramics and ceramic-matrix composites. Note that all mechanisms are extrinsic in nature and in addition to shielding the crack tip, they also serve to promote inelastic deformation which results in a nonlinear stress/strain relationship [25]. *Reproduced with permission.*

as they operate at, or primarily behind, the crack tip to lessen the effective crack-driving force; they are therefore mechanisms of crack-growth toughening. Indeed, it is the early portion of the R-curve (over the first 100 micrometers or so of crack extension) that is also very important for ceramics as this governs their fracture strength at realistic flaw sizes. Consequently, the resulting strengths depend markedly on the details of the R-curve and of course the initial flaw sizes, such that optimizing strength versus toughness can involve different choices of microstructures.

Crack deflection: As described in Section 3.6.2, deflecting a mode I crack off the plane of maximum tensile stress acts to reduce the effective (local) driving force at the crack tip. This has been shown to play a significant role in governing the toughness and subcritical crack growth in a wide range of materials, from the toughness of brittle ceramics [26] to fatigue-crack growth in ductile metallic alloys [27]. For the simply tilted crack in Fig. 3.20, a reduction in crack-driving force for in-plane deflections of, say, 30 and 90 degrees, can be on the order of 10% and 50%, respectively (as per Eqs. 3.19 and 3.20); much larger reductions can be achieved with associated twisting of the crack. This reduction in driving force depends solely on the angle of the deflection, but the size of the deflected crack segment can have an additional effect on the kinetics of crack advance, as it dictates the extent of crack extension which experiences the reduced driving force. Accordingly, shielding through crack deflection can be more significant in affecting subcritical crack-growth rates, particularly in fatigue where it can promote additional crack-tip shielding via crack closure.

Microcrack and transformation toughening: A particularly potent form of extrinsic toughening in ceramics is transformation toughening, which is seen in many zirconia-based ceramics such as partially-stabilized zirconia. Here, an in situ stress-induced martensitic phase transformation can occur in the high stress region ahead of the crack tip where tetragonal ZrO_2 precipitates transform to a monoclinic phase with an incumbent dilation of $\varepsilon_T \approx 2\%-4\%$. Due to the surrounding constraint of untransformed material, this generates a compressive stress of $\sigma_R \sim E' f_v \varepsilon_T$ within the transformation zone at the crack tip, where E' is the elastic modulus and $f_v \sim 0.2$ is the volume fraction of transforming precipitates. This results in a compressive shielding stress intensity, K_s, opposing the applied K_I, such that as the crack extends into this zone, the effective stress intensity at the crack tip is progressively reduced, which leads to a rising R-curve where the peak (steady-state) K_c toughness is given by Ref. [28]:

$$K_c = K_o + 0.22 \, \varepsilon_T \, E' \, f_v \, \sqrt{h} \ . \tag{6.10}$$

Here h is the size of the transformation zone, $h \sim \frac{1}{2\pi}(K_c/\sigma_T)^2$, where σ_T is the transformation stress. The fracture toughness K_c for partially stabilized zirconia can approach 15 MPa√m, a factor of some five times larger than the intrinsic toughness K_o of non-transforming zirconia.

Although a highly effective extrinsic toughening mechanism, it is only applicable to specific ceramics at the relatively low temperatures where the transformation is thermodynamically viable. Constrained microcrack toughening, however, which is a similar mechanism, can in principle occur at any temperature. Many brittle materials are known to microcrack, including rocks, some ceramics and bone. The occurrence of constrained microcracking in the high stresses in the vicinity of a (macro) crack can in principle result in a dilation effect similar to transformation-toughening, but the magnitude of this extrinsic toughening effect does not appear to be that significant. However, microcracking is a form of inelasticity which can relieve regions of high stress, and in many materials it can also lead to uncracked ligament bridging, which can generate significant toughening in many engineering and biological materials.

Crack bridging: The most omnipresent form of shielding is crack bridging, which is found in different forms in almost all classes of materials, particularly natural and biological materials. Here intact "features" span the crack as it opens thereby carrying load that would be otherwise used to extend the crack. These "features" include certain reinforcements, e.g., fibers in polymeric and ceramic composites, uncracked ligaments between a main crack and (micro) cracks initiated ahead of it, which is a common toughening mechanism in ceramics, rocks, and biological materials like bone and tooth dentin [29], and frictional bridging associated with interlocking grains during intergranular fracture, which is the primary and most potent source of toughening in monolithic structural ceramics with engineered grain boundaries, e.g., Al_2O_3, Si_3N_4, and SiC. Ductile phases can also act as deformable bridges, such as in rubber-toughened polymers and certain ductile-phase intermetallic composites.

6.4.3 Toughening in polymeric materials

Polymers do not contain crystallographic planes, dislocations, or grain boundaries but rather consist of (generally) covalently bonded molecular networks, which in thermoplastic polymers are in the form of long chains held loosely together by secondary van der Waals bonds. Ultimate fracture normally requires breaking the primary bonds, but the secondary bonds often play a major role in the deformation mechanisms that lead to fracture. As they are generally rate-dependent materials, factors such as strain rate, temperature, and molecular structure have a strong influence on ductility and toughness in polymers. At high rates/ low temperatures (relative to the glass-transition temperature), polymers tend to be brittle, as there is insufficient time for yielding or larger-scale viscoelastic mechanisms to respond to stress. Shear yielding and crazing are competing mechanisms here. Shear yielding resembles plastic flow in metals: molecules slide with respect to one another when subjected to critical shear stress. Crazing, which occurs in glassy polymers subjected to tensile stresses, represents highly localized deformation that leads to cavitation (void formation). Compared with shear yielding, crazes are more likely ahead of a crack tip because of the triaxial stress-state there. The crack advances when the fibrils at the trailing edge of the craze rupture. In other words, cavities in the craze zone coalesce with the crack tip, similar to microvoid coalescence in metals. Craze crack growth can either be stable or unstable, depending on the relative toughness of the material.

6.4.4 Toughening in composite materials

The incorporation of reinforcements in the form of fibers, whiskers, or particles can also toughen materials, although the motivation may be rather to increase strength and/or stiffness. For toughening, crack bridging is again the most prominent mechanism, particularly in *ceramic-matrix composites*; by utilizing fibers with weak fiber/matrix bonding, when the matrix fails, the fibers are left intact spanning the crack wake and can act as bridges to inhibit crack opening [24].

Analogous toughening in *metal-matrix composites* is considerably less advanced, in part because many such composites are designed with strong reinforcement-matrix interfaces and thus do not develop crack bridging to any significant degree. In metal-matrix composites discontinuously reinforced with a brittle particulate phase, such as aluminum alloy-SiC composites, the intent primarily is to increase the strength and sometimes wear resistance. However, toughness can be generated by crack bridging which results from the uncracked ligaments created where microcracks, formed at SiC particles some distance ahead of the crack tip, have yet to link to the main crack.

In continuous-fiber-reinforced *polymer composites,* which are expensive yet seeing ever increasing use in aerospace structures and in other lightweight structural applications, high volume fractions of carbon (graphite) fibers, with strong matrix/fiber interfaces, are added for strength and stiffness; to a lesser degree, continuous glass and ceramic fibers have also been used for improved strength. For discontinuous reinforcements, additions of rubber particles to polymers can promote both crack deflection and bridging, as the crack will tend to follow the low modulus rubber phase which, while it remains intact, can act as a deformable bridge across the crack.

References

[1] R.O. Ritchie, J.F. Knott, J.R. Rice, Relationship between critical tensile stress and fracture toughness in mild steel, J. Mech. Phys. Solid. 21 (1973) 395.
[2] F.A. McClintock, A criterion for ductile fracture by the growth of holes, J. Appl. Mech. 35 (1968) 363.
[3] A.C. Mackenzie, J.W. Hancock, D.K. Brown, On the influence of state of stress on ductile failure initiation in high strength steels, Eng. Fract. Mech. 9 (1977) 167.
[4] J.R. Rice, G.F. Rosengren, Plane strain deformation near a crack tip in a power-law hardening material, J. Mech. Phys. Solid. 16 (1968) 1.
[5] J.W. Hutchinson, Singular behavior at the end of a tensile crack tip in a hardening material, J. Mech. Phys. Solid. 16 (1968) 13.
[6] D.M. Tracey, Finite element solutions for crack-tip behavior in small-scale yielding, J. Eng. Matls. Tech. 98 (1976) 146.

[7] J.R. Rice, M.A. Johnson, The role of large crack tip geometry changes in plane strain fracture, in: M.F. Kanninen, W.F. Adler, A.R. Rosenfield, R.I. Jaffee (Eds.), Inelastic Behavior of Solids, McGraw-Hill, New York, NY, 1970, p. 641.

[8] R.M. McMeeking, Finite deformation analysis of crack tip opening in elastic-plastic materials and implications for fracture initiation, J. Mech. Phys. Solid. 25 (1977) 357.

[9] R.O. Ritchie, A.W. Thompson, On the macroscopic and microscopic analyses for crack initiation and crack growth toughness in ductile alloys, Metall. Trans. A. 16A (1985) 233.

[10] C.F. Shih, Division of Engineering, Brown Univ. Rep. no. MRL E-147, Providence, RI, 1983.

[11] R.O. Ritchie, The conflicts between strength and toughness, Nat. Mater. 10 (2011) 817.

[12] T. Lin, A.G. Evans, R.O. Ritchie, Statistical model of brittle fracture by transgranular cleavage, J. Mech. Phys. Solid. 34 (1986) 477.

[13] J. Watanabe, T. Iwadate, Y. Tanaka, T. Yokobori, K. Ando, Fracture toughness in the transition region, Eng. Fract. Mech. 28 (1987) 589.

[14] R.O. Ritchie, W.L. Server, R.A. Wullaert, Critical fracture stress and fracture strain models for the prediction of lower and upper shelf toughness in nuclear pressure vessel steels, Metall. Trans. A. 10A (1979) 1557.

[15] D.A. Curry, Predicting the temperature and strain rate dependence of the cleavage fracture toughness of ferritic steels, Mater. Sci. Eng. 43 (1980) 135.

[16] D.M. Parks, Interpretation of irradiation effects on fracture toughness of a pressure-vessel steel in terms of crack tip stress, J. Eng. Matls. Tech. 98 (1976) 30.

[17] P. Novak, R. Yuan, B.P. Somerday, P. Sofronis, R.O. Ritchie, A statistical, physical-based, micro-mechanical model of hydrogen-induced intergranular fracture in steel, J. Mech. Phys. Solid. 58 (2010) 206.

[18] D.A. Curry, J.F. Knott, Effect of microstructure on cleavage fracture toughness in mild steel, Met. Sci. 13 (1979) 341.

[19] J.R. Rice, D.M. Tracey, On the ductile enlargement of voids in triaxial stress fields, J. Mech. Phys. Solid. 17 (1969) 201.

[20] J.R. Rice, W.J. Drugan, T.-L. Sham, Elastic-plastic analysis of growing cracks, in: ASTM STP 700, American Society for Testing and Materials, West Conshohocken, PA, 1980, p. 189.

[21] J.F. Knott, Fundamentals of Fracture Mechanics, Butterworths, London, UK, 1973.

[22] M.E. Launey, R.O. Ritchie, On the fracture toughness of advanced materials, Adv. Mater. 21 (2009) 2103.

[23] B. Gludovatz, A. Hohenwarter, D. Cartoor, E.H. Chang, E.P. George, R.O. Ritchie, A fracture resistant high-entropy alloy for cryogenic applications, Science 345 (2014) 1153.

[24] A.G. Evans, Perspective on the development of high toughness ceramics, J. Am. Ceram. Soc. 73 (1990) 187.

[25] R.O. Ritchie, Mechanisms of fatigue-crack propagation in ductile and brittle solids, Int. J. Fract. 100 (1999) 55.

[26] K.T. Faber, A.G. Evans, Crack deflection processes — I. Theory, Acta Metall. 31 (1983) 565.

[27] S. Suresh, Crack deflection: implications for the growth of long and short fatigue cracks, Metall. Trans. A 14A (1983) 2375.

[28] R.M. McMeeking, A.G. Evans, Mechanics of transformation-toughening in brittle materials, J. Am. Ceram. Soc. 65 (1982) 242.

[29] U.G.K. Wegst, H. Bai, E. Saiz, A.P. Tomsia, R.O. Ritchie, Bioinspired structural materials, Nat. Mater. 14 (2015) 23.

7

Application to subcritical crack growth

7.1 Introduction

In 1961, Paul Paris, along with colleagues Mario Gomez and Bill Anderson at Boeing, made the proposition that subcritical crack growth by fatigue could also be correlated to the stress intensity [1]. The premise was that as K_I uniquely characterizes the local stresses and displacements at a crack tip, then the crack growth increment per cycle, da/dN, should also be correlated to current K_I value. In a paper that was apparently rejected by three well-known journals before eventually appearing in the University of Washington alumni magazine *The Trend in Engineering*, they proposed that da/dN should be proportional to K_{max}^4. This was not quite right but Paris with Erdogan published two years later what is now known as the Paris law for fatigue-crack propagation as [2]:

$$\frac{da}{dN} = C\Delta K^m,$$

(7.1)

where da/dN is proportional to the stress-intensity range, ΔK, i.e., to the difference between the maximum and minimum stress intensities in a given cycle ($\Delta K = K_{max} - K_{min}$), and C and m (which is on the order of 2–4 for most metals) are material scaling constants.

It was soon realized, specifically by Johnson and Paris in 1967 [3], that such correlations could be developed for other forms of subcritical cracking, which is defined as crack extension at stress intensities (or an equivalent crack-driving force) which are less than that required for outright failure, e.g., prior to $K = K_c$. For example, such crack-growth relationships can be used for time-dependent cracking (da/dt), such as

101

environmentally-assisted crack growth by stress corrosion or hydrogen-assisted fracture, where relationships of the form of:

$$\frac{da}{dt} = C' K_I^{n'} ,$$ (7.2)

can be used; C' and n' are again scaling constants. This is described in Section 7.2.

In addition, there has been extensive work on subcritical creep-crack growth using characterizing parameters for the various constitutive behavior associated with creep deformation that can be loosely described as rate-dependent versions of J, as described below in Section 7.3.

In addition to providing a means to characterize subcritical crack growth behavior, the use of fracture mechanics in this context presents an alternative approach to predicting fatigue, environmentally-assisted fracture, and creep failures, using so-called damage-tolerant (or defect-tolerant) life-prediction strategies. Instead of using traditional stress- (or strain-) based total life approaches, where lifetimes are estimated in terms of the time, or number of loading cycles, both to initiate and propagate a crack to cause failure in a laboratory (often smooth-bar) specimen (a good example is the use of S-N (Wöhler) curves for fatigue), the damage-tolerant life-prediction strategies recognize that all structures are flawed, and that cracks may initiate early in service life and propagate subcritically. Lifetimes are then assessed on the basis of the time or number of cycles for the largest undetected crack to grow to failure, as might be defined by an allowable strain, or limit load, or $K_I = K_c$ criterion (described in Section 7.5). Implicit in such analyses is that subcritical crack growth can be characterized in terms of a characterizing parameter that describes local conditions at the crack tip, yet may be determined in terms of loading parameters, crack size, and geometry, to yield crack-growth relationships of the form of Eqs. (7.1) and (7.2). By assuming that flaws pre-exist in a structure, i.e., that the crack-initiation life is effectively zero, the crack-propagation life can be predicted by the integration of such relationships, from some initial crack size to a final crack size defining failure, in order to develop a conservative estimate of the safe lifetime.

In this chapter, we will briefly describe the application of fracture mechanics to time-dependent cracking but, in the interests of brevity, we will devote the majority of the chapter to fatigue, which is widely considered to be the most prevalent mode of *in service* fractures.

7.2 Environmentally-assisted cracking

There are many processes of time-dependent environmentally-assisted cracking, but the most notable are stress-corrosion cracking (SCC), which results from the conjoint action of stress and a corrosive aqueous environment,[1] and hydrogen-assisted cracking, which results from stress in the presence of hydrogen-containing or hydrogen-producing environments. There are innumerable mechanisms [e.g., 4,5] by which these processes have been proposed to occur, but they can be broadly separated into two basic classes: anodic- or active-path corrosion and hydrogen embrittlement mechanisms (Fig. 7.1).

With respect to aqueous stress-corrosion, the environmental mechanism contributing to the fracture process can be controlled by anodic processes at the crack tip (active-path corrosion) where the metal is oxidized, or by cathodic processes where reduction reactions at the crack tip can release hydrogen which enters the metal and cause deleterious effects (hydrogen embrittlement). Hydrogen embrittlement can also occur in a hydrogen gas (or hydrogen-producing environment such as H_2S), where hydrogen can chemisorb on the fresh (active) crack surface before diffusing into the metal, aided by the stress gradient, to collect in traps, such as dislocations and interfaces. As hydrogen is an interstitial element, it can diffuse rapidly in the metal; the resulting primary forms of embrittlement are premature decohesion at interfaces such as grain boundaries, interactions with dislocations which affect their motion and induce local plastic instabilities, or from the formation of brittle hydrides in certain materials, such as Ti and Zr.

In the presence of a stress, these mechanisms can cause subcritical cracking where the growth rate, da/dt, is a power-law function of the stress intensity, K_I, as per Eq. (7.2). However, the actual form of the da/dt vs. K_I relationship, represented on log-log (or log-linear) plots and often called v-K (velocity-stress intensity) curves, generally resembles a sigmoidal curve in form which can be classified into three main regimes, I, II, and III (Fig. 7.2).

Although different material/environmental combinations can display various types of behavior, these three regions in the v-K curves often have similar specific origins. In region I, just above the threshold stress intensity for environmentally-assisted cracking, termed K_{Iscc} or K_{TH}, the crack-opening displacements can create fresh surface at the crack tip where the environmental species can adsorb or cause corrosion, leading to initially rapid cracking. The growth rates in this region are a function of the nature of the environment (and temperature T and pressure p) and a very strong function of K_I. Consequently, as the slope of the plot, i.e., the exponent n' in Eq. (7.2), can become very high, it is often better to express the da/dt vs. K_I relationship in exponential form, e.g.,

[1] For example, window-pane glass will crack subcritically at a stress intensity of about one-third of its toughness due to SCC. Indeed, you can apparently cut glass with a pair of scissors under water - but please do not try this at home!

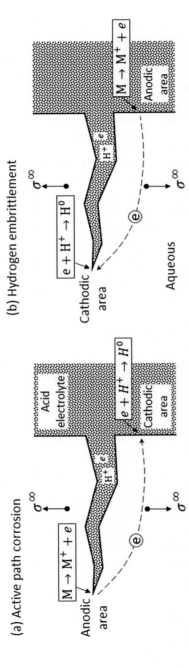

FIGURE 7.1 Two major classes of mechanisms of stress-corrosion cracking in aqueous environments, in this example an acid electrolyte. In (a) *active-path corrosion* mechanisms, the primary environmental contribution to cracking at the crack tip is an anodic process, e.g., involving metal oxidation ($M \rightarrow M^+ + e$), whereas in (b) *hydrogen embrittlement* mechanisms, the primary environmental effect at the crack tip is cathodic, e.g., a reduction reaction to release hydrogen, which then enters the material to cause deleterious effects through decohesion at interfaces, affecting dislocation motion to promote plastic instability, or in certain materials through brittle hydride formation.

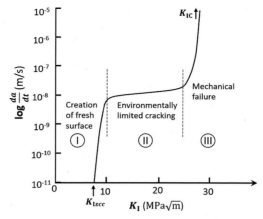

FIGURE 7.2 Schematic of a v-K curve for environmentally-assisted cracking showing the variation in crack-growth rate, da/dt, as a function of the stress intensity, K_I. Such plots tend to be sigmoidal in shape, between a threshold stress intensity, i.e., K_{Iscc} for SCC or K_{TH} for other forms of subcritical cracking, and instability, i.e., at K_{Ic} or the limit load. Region I, where the exponent n' in Eq. (7.2) can be very high, is often associated with the creation of fresh surface at the crack tip, which can lead to an initial burst of cracking. In region II, often $n' \to 0$ as steady-state cracking can become rate-limited by some environmental process, such as chemisorption at the crack tip or hydrogen transport in the lattice. Region III is essentially governed by mechanical failure at instability, e.g., as $K_I \to K_{Ic}$.

$(da/dt)_I = f\ (K_I,\ T,\ p,\ \text{environment}) = C_1 \exp(n_1 K_I)$, where C_1 and n_1 are material constants. In the steady-state region II, the exponent n' can approach zero, i.e., da/dt can become almost independent of K_I, as the process of crack growth is restricted by a non-mechanical rate-limiting step, such as hydrogen transport or chemisorption of an active species at the crack tip; here $(da/dt)_{II} = f\ (T,\ p,\ \text{environment})$. In region III where $K_I \to K_{Ic}$ (or whatever form of instability is relevant), cracking is largely associated with mechanical failure.

Actual v-K data are shown in Fig. 7.3 for two ultrahigh-strength NiCrMo steels, namely quenched and tempered 4340 and Si-modified 4340 (300-M), tested in distilled water at 23°C [6]. Several points are particularly noteworthy here. Firstly, there are distinct differences in the regime II growth-rate plateaus which here are related to a rate-limiting step associated with the transport of hydrogen in the lattice. If we compare the two 300-M microstructures, both of which have yield strengths of 1500 MPa, the one marked "870°C, oil, iso 250°C" was quenched from 870°C and held at 250°C to create 10%–12% retained austenite in its martensitic structure, compared to the "870°C, oil" structure that was quenched directly in oil to form martensite with <2% austenite. It is apparent that microstructure with the lower plateau has the larger volume fraction of retained austenite which forms around the

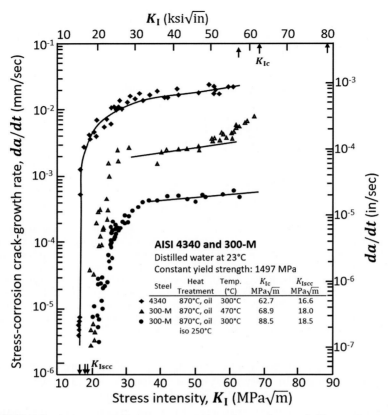

FIGURE 7.3   Variation in the stress-corrosion crack-growth rate, da/dt, as a function of the stress intensity K_I, for two ultrahigh-strength NiCrMo steels, quenched and tempered 4340 and Si-modified 4340 (300-M), tested in distilled water at 23°C. Note that although the K_{Ic} toughness of these steels (in their respective tempered conditions) ranges from ~63 to 89 MPa$\sqrt{m}$, they start to subcritically crack in water at K_{Iscc} K_I levels between 15 and 20 MPa$\sqrt{m}$ [6]. *Reproduced with permission.*

martensitic laths; as the austenite is face-centered cubic, hydrogen diffusion is some two orders of magnitude slower than in the less close-packed body-centered martensitic structure, which acts to slow down hydrogen transport to the crack tip, and reduce the steady-state growth rates. A second notable point is that although the fracture toughness, K_{Ic}, of these steels (in their respective tempered conditions) range from ~63 to 89 MPa$\sqrt{m}$, they start to subcritically crack in water at K_{Iscc} levels a factor of four lower, i.e., between 15 and 20 MPa$\sqrt{m}$. Indeed, such ultrahigh-strength steels are particularly susceptible to hydrogen-assisted cracking in hydrogen gas and aqueous environments.

TABLE 7.1 Summary of the typical K_{TH} or K_{Iscc} thresholds for environmentally-assisted cracking of a series of metallic alloys, in terms of their yield strength and K_{Ic} fracture toughness values.

Material	Environment	Yield strength σ_y	Fracture toughness, K_{Ic}	Environment threshold K_{Iscc} or K_{TH}
		(MPa)	(MPa√m)	(MPa√m)
4340 steel (200°C temper)	Salt water	1600	60	17
	H$_2$ gas			18
	H$_2$S gas			16
(600°C temper)	Salt water	1000	~150	120
18Ni (180) Maraging steel	NaCl solution	1200	130	120
18Ni (250) Maraging steel		1700	89	14
2024-T3 Al-Cu alloy	NaCl solution	370	35	11
7075-T6 Al-Zn-Mg alloy	NaCl solution	505	28	20
Ti-8Al-1Mo-1V alloy	Water	855	100	29
	Methanol			15

A brief listing of various metal alloys subjected to different environments is shown in Table 7.1, where it is clear that the higher strength microstructures are often the most prone to environmentally-assisted cracking, as defined by their low threshold K_{TH} values compared to K_{Ic}, although some materials are highly susceptible due to their composition or specific microstructural features, such as precipitate-free zones at grain boundaries in certain aluminum alloys.

7.3 Creep-crack growth

7.3.1 C* integral

As noted above, there have been extensive studies on the application of fracture mechanics to describe creep-crack growth in the form of characterizing parameters defined under various mode of creep deformation. The first of these parameters was defined for secondary (steady-state) creep, where the constitutive law can be expressed in terms of the strain rate $\dot{\varepsilon}_{ij}$ as a power-law function of stress, σ_{ij} by $\dot{\varepsilon}_{ij} = A\sigma_{ij}^{n_c}$, where n_c is the creep exponent and A is a material constant. This parameter was termed C* and was defined as a path-independent integral in the form of the definition of J in Eq. (4.3) only with strain rates replacing strains, displacement rates replacing displacements, and stresses staying the same [7,8]. Thus, for any anticlockwise path Γ taken around the tip of a crack linking the lower to the upper crack surface, as in Fig. 4.2, the C^* integral can be defined as:

$$C^* = \int_\Gamma \left(\dot{W}_{se} dy - T_i \frac{\partial \dot{u}_i}{\partial x} ds \right), \qquad (7.3)$$

where T_i is the traction vector defined by the outward normal n_j ($T_i = \sigma_{ij} n_j$) along the contour Γ, $\dot{u}_i$ is the displacement rate vector, ds is a length increment, and $\dot{W}_{se}$ is the strain energy rate density given by:

$$\dot{W}_{se} = \int_0^{\dot{\varepsilon}_{ij}} \sigma_{ij} \, d\dot{\varepsilon}_{ij}. \qquad (7.4)$$

Further by analogy to $J = \frac{\eta}{Bb} \int_0^\Delta P d\Delta$, where $Pd\Delta$ is the energy absorbed by the specimen, Bb is the area of the ligament, i.e., the specimen thickness times the uncracked ligament length, and η is a geometry factor (Eqs. 4.14 and 4.15), C^* can also be defined in energy, actually energy rate, terms as:

$$C^* = \frac{\eta}{Bb} \int_0^{\dot{\Delta}} P d\dot{\Delta}, \qquad (7.5)$$

where P is the applied load and $\dot{\Delta}$ is the load-line displacement rate.

Finally, C^* can be identified as the characterizing parameter for an HRR field, analogous to Eq. (4.2), but now based on a steady-state creep constitutive law $\dot{\varepsilon}_{ij} = A\,\sigma_{ij}^{n_c}$, where the crack-tip stress and strain rates as a function of r from the crack tip are given by:

$$\sigma_{ij} = \left[\frac{C^*}{I_N A r} \right]^{1/(n_c+1)} \tilde{\sigma}_{ij} (\theta, n_c), \qquad (7.6a)$$

$$\dot{\varepsilon}_{ij} = \left[\frac{C^*}{I_N A r} \right]^{n_c/(n_c+1)} \tilde{\varepsilon}_{ij} (\theta, n_c). \qquad (7.6b)$$

Here, the strain-hardening rate N in the J definition in Eq. (4.2) is replaced by the creep exponent n_c, and I_N, $\tilde{\sigma}_{ij}(\theta, n_c)$ and $\tilde{\varepsilon}_{ij}(\theta, n_c)$ have the same values as those in Eq. (4.2). Hence, as C^* can be identified as a singular parameter governing the unique crack-tip stress and strain fields under extensive steady-state creep deformation, we can correlate the rate of growth of a creep crack $(da/dt)_c$ with C^*, with relationships of the form:

$$\left(\frac{da}{dt} \right)_c = \chi_c (C^*)^{m_c}, \qquad (7.7)$$

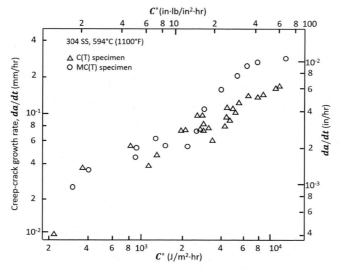

FIGURE 7.4 Creep-crack growth rates, da/dt, as a function of the C^*-integral for an austenitic 304 stainless steel, tested at 594°C with C(T) and MC(T) specimens [9]. *Reproduced with permission.*

where χ_c and m_c are material scaling constants with the exponent $m_c = n_c/(n_c + 1)$. This power-law relationship has been found to well represent the creep-crack growth rates in many metallic materials under conditions of extensive steady-state creep. This is shown in Fig. 7.4, where creep-crack growth rates in C(T) and MC(T) specimens of a 304 austenitic stainless steel at 594°C are plotted as a function of C^* [9].

7.3.2 C(t) and C_t parameters

However, when the creep field initially forms under conditions of small-scale creep, the creep zone stress fields are governed by another characterizing parameter $C(t)$ [10,11] which varies with time but becomes equal to C^* in the long-term limit of steady-state creep. $C(t)$ can be defined by a path-independent integral, identical to Eq. (7.3), only for contour taken very close to the crack tip where the creep strains exceed the elastic strains. Furthermore, it can interpreted as the governing parameter in an HRR description of the small-scale creep stress and strain-rate fields, e.g.,

$$\sigma_{ij} = \left[\frac{C(t)}{I_N A r} \right]^{1/(n_c+1)} \widetilde{\sigma}_{ij}(\theta, n_c) . \tag{7.8}$$

$C(t)$ is an appealing characterizing parameter for correlating creep-crack growth rates for conditions of small-scale creep and its transition to extensive steady-state creep, but it is difficult to measure experimentally under small-scale and transition creep conditions. Accordingly, Saxena [12] proposed an analogous parameter, C_t, that could still be used to define creep-crack growth over this wide range of creep conditions but was easier to experimentally determine. C_t is defined, as was C^* in Eq. (7.5), as the creep component of the power release rate, but with an additional component for small-scale creep, $(C_t)_{\text{ssc}}$, modeled by analogy to the K_I-fields, such that:

$$C_t = (C_t)_{\text{ssc}} + C^* . \tag{7.9}$$

Indeed, experimental studies [13] have shown that C_t correlates creep-crack growth rates far better than K_I or C^* as a single parameter for a wide range of conditions from small-scale to extensive steady-state creep conditions.

Finally, it is of note that there are other characterizing parameters that have been proposed to characterize the various other stages of creep cracking, in particular for primary creep. The interested reader is referred to Ref. [14] for further details.

7.4 Fatigue-crack growth

Cyclic fatigue involves the microstructural damage and failure of materials under cyclically varying loads. Structural materials, however, are seldom designed for fatigue resistance; rather metallic alloys are generally designed for strength, intermetallics for ductility, and ceramics for toughness, yet, in engineering service, their structural integrity is most often limited by their mechanical performance under cyclic loads. In fact, it is generally considered that over 80% of all service failures can be traced to mechanical fatigue, whether in association with cyclic plasticity, sliding, or physical contact (fretting, rolling, contact fatigue), environmental damage (corrosion fatigue) or elevated temperatures (creep fatigue); despite this, mechanistically it is the least understood mode of fracture.

The introduction of the Paris law [2] (Eq. 6.1) in the early 1960s ignited a flurry of interest in the propagation of fatigue cracks. However, akin to environmentally-assisted cracking, the full relationship between the fatigue-crack growth rate and the stress intensity is sigmoidal in shape. Typical da/dN vs. ΔK data in the form of a log-log plot for a high-strength 300-M alloy steel, where the strength level has been changed by heat treatment, is shown in Fig. 7.5 [15]. Although ΔK is the primary governing parameter in metal fatigue, in actuality, growth rates depend upon several other factors, i.e., $da/dN = f$ (ΔK, K_{max} (or $R = K_{\text{min}}/K_{\text{max}}$), frequency, environment, wave form, etc).

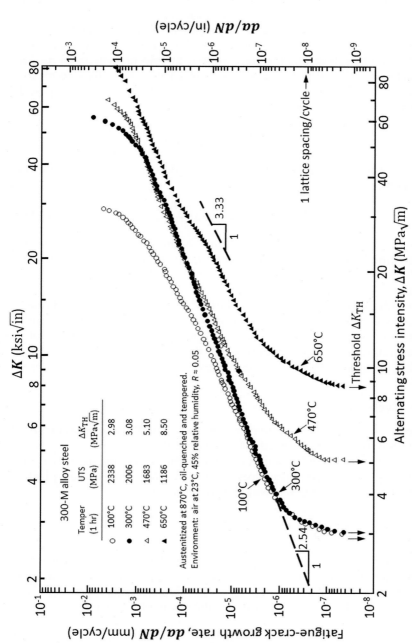

FIGURE 7.5 Typical sigmoidal variation in fatigue-crack growth rates da/dN, as a function of the stress-intensity range ΔK, for a quenched and tempered 300-M alloy steel at ambient temperature, where the tensile strength has been varied by changing the tempering temperature. Tests performed at a low R-ratio = K_{min}/K_{max} of 0.05 [15]. From R.O. Ritchie, *Influence of microstructure on near-threshold fatigue crack propagation in ultra-high strength steel, Metal Sci. 11 (1977) 368; copyright ©Institute of Materials, Minerals and Mining, reprinted by permission of Informa UK Limited, trading as Taylor & Francis Group, www.tandfonline.com on behalf of Institute of Materials, Minerals and Mining.*

From this plot in Fig. 7.5, it is apparent that the simple Paris law (Eq. 7.1) is generally a good fit to the data in the mid-range of growth rates. However, numerous laws have been proposed to account for the acceleration in growth rates as K_{max} approaches fracture instability, e.g., at K_c, at high growth rates, and the deceleration at low growth rates as ΔK approaches the so-called threshold stress intensity ΔK_{TH}. To account for the high growth rate regime, one well-known relationship was proposed by Forman et al. [16]:

$$\frac{da}{dN} = \frac{C_1\,\Delta K^m}{(1-R)K_c - \Delta K}\,, \tag{7.10}$$

whereas to account for the near-threshold behavior, Klesnil and Lukas [17] suggested:

$$\frac{da}{dN} = C_2\left(\Delta K^m - \Delta K_{TH}^m\right), \tag{7.11}$$

where C_1, C_2, and m are scaling constants. Naturally, these relationships can be combined to fit the entire curve; indeed, all these crack-growth "laws" are essentially empirical curve-fits to the data based on the original Paris power-law relationship.

In addition to the $da/dN \propto \Delta K^m$ approach, other fracture mechanics power-law expressions have been proposed to characterize fatigue-crack growth in terms of the range of crack-tip opening displacement (ΔCTOD) [18], or under large-scale yielding by the cyclic J-integral (ΔJ) [19], although, compared to the Paris law, these approaches are used rarely.

7.4.1 Cyclic plastic zone

Before we discuss the different regimes of growth-rate behavior that are apparent in Fig. 7.5, we need to briefly describe the nature of the cyclic plastic zone at the crack tip [20]. As shown by the schematic in Fig. 7.6, on first loading to the maximum stress intensity, K_{max}, a maximum plastic zone is formed in the regular way described in Section 3.2:

$$r_{y,max} \sim \frac{1}{2\pi}\left(\frac{K_{max}}{\sigma_y}\right)^2, \tag{7.12}$$

where σ_y is the yield stress. However, on unloading to K_{min}, we need to superimpose an elastic unloading stress of $-2\sigma_y$ on the material as the crack tip is stressed from $+\sigma_y$ to $-\sigma_y$, which results in a zone of compression at the crack tip, known as the cyclic plastic zone:

$$r_{y,c} \sim \frac{1}{2\pi}\left(\frac{\Delta K}{2\sigma_y}\right)^2, \tag{7.13}$$

which is ~¼ of the maximum plastic zone size for zero-tension ($R = 0$) loading (σ_y in Eq. 7.13 is strictly the cyclic yield stress).

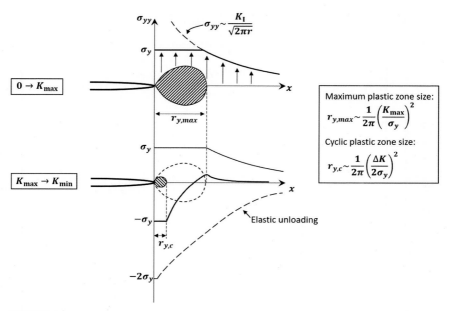

FIGURE 7.6 Formation of the maximum and cyclic plastic zones at the tip of a fatigue crack being cycled by loading to K_{max} and then unloading to K_{min}.

As with monotonic loading, provided these plastic zones are embedded within the K-field, the stress intensity will still uniquely define the crack-tip stresses and strains.

7.4.2 Regimes of crack-growth behavior

Returning to the da/dN vs. ΔK plot in Fig. 7.5, there are clearly three regimes of crack-growth behavior, which are summarized with their respective behavioral properties in the schematic diagram in Fig. 7.7 [15].

Specifically, results of fatigue-crack growth rate tests for most ductile materials display the following characteristics:

(i) near-threshold region A at low values of ΔK and da/dN (less than $\sim 10^{-9}$ m/cycle) approaching the fatigue threshold, ΔK_{TH}, below which fatigue cracks appear to be dormant, or grow at vanishingly small growth rates.

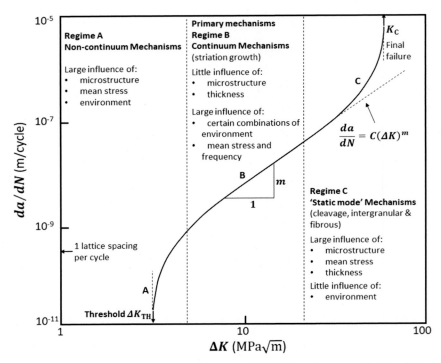

FIGURE 7.7 Schematic illustration of the typical variation in fatigue-crack growth rates *da/dN* as a function of the applied stress-intensity range ΔK in metallic materials, showing the regimes of primary growth-rate mechanisms and effects of several major variables on crack-growth behavior [15]. *From R.O. Ritchie. Influence of microstructure on near-threshold fatigue crack propagation in ultra-high strength steel, Metal Sci. 11 (1977) 368; copyright ©Institute of Materials, Minerals and Mining, reprinted by permission of Informa UK Limited, trading as Taylor & Francis Group, www.tandfonline.com on behalf of Institute of Materials, Minerals and Mining.*

(ii) an intermediate region B ($\sim 10^{-9}$ to 10^{-6} m/cycle) of power-law behavior described by the Paris equation (Eq. 7.1).

(iii) upper region C of accelerating crack growth (above $\sim 10^{-6}$ m/ cycle) as K_{max} approaches K_c or gross plastic deformation of the specimen, e.g., at the limit load.

We will briefly describe the characteristics of each regime in reverse order.

High growth-rate regime: As the K_{max} value during the fatigue cycle approaches the fracture instability, e.g., at K_c or the limit load, the growth rates can exceed those based on an extrapolation of the Paris law in the intermediate growth-rate regime (region C). This results from the

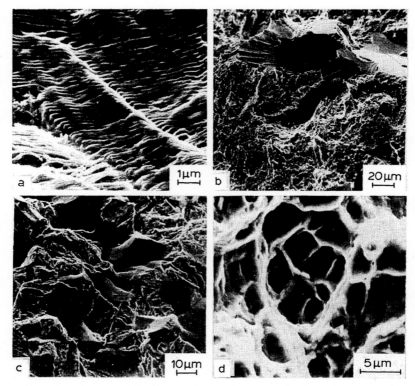

FIGURE 7.8 Scanning electron microscopy images of fatigue-crack propagation at intermediate (region B) and high (region C) growth rates in steels in room air at $R = 0.1$, showing (a) ductile striations in a 9Ni-4Co steel at $\Delta K = 30$ MPa$\sqrt{m}$, (b) additional cleavage cracking in mild steel at $\Delta K = 40$ MPa$\sqrt{m}$, (c) additional intergranular cracking in 4Ni-1½Cr steel at $\Delta K = 40$ MPa$\sqrt{m}$, and (d) microvoid coalescence in the 9Ni-4Co steel at $\Delta K = 70$ MPa$\sqrt{m}$ [23]. *From R.O. Ritchie. Near-threshold fatigue crack propagation in steels, Int. Metals Rev. 24 (1979) 205; copyright ©Institute of Materials, Minerals and Mining, reprinted by permission of Informa UK Limited, trading as Taylor & Francis Group, www.tandfonline.com on behalf of Institute of Materials, Minerals and Mining.*

occurrence of additional monotonic modes of fracture, called "static modes" [21] (Fig. 7.8), principally microvoid coalescence, cleavage, and intergranular cracking, to the cyclic crack advance (striation) mechanism [22], which serves to accelerate fatigue-crack growth rates until the onset of overload fracture. Because these monotonic fracture modes are governed by K_{max}, crack growth is increased at higher R-ratios, but cracking is relatively unaffected by the nature of the environments in this regime because the growth rates are generally too fast.

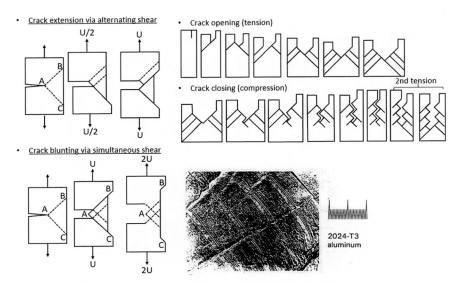

FIGURE 7.9 Mechanism of striation formation associated with alternating crack-tip blunting and resharpening via a process of alternating or simultaneous shear, i.e., dislocation emission and partial reversal, at the crack tip. U is the applied displacement perpendicular to the crack in each cycle [24,25]. *Reproduced with permission.*

Intermediate growth-rate regime: In this region B, between typically ~10^{-9} to 10^{-6} m/cycle, the Paris law generally applies, and crack advance commonly occurs by the classical (intrinsic) fatigue mechanism of ductile striations [22] (Fig. 7.8a). These striations represent the local cyclic crack-growth increment each cycle, and are formed within the cyclic plastic zone by crack-tip blunting on the loading cycle with subsequent resharpening of the crack tip on unloading (Fig. 7.9).

One interpretation of the striation crack-growth process is crack blunting, i.e., dislocation emission from the crack-tip, by alternating shear or simultaneous shear [23,24], the latter process occurring if strain hardening on one shear band facilitates shear on the other one (Fig. 7.9). New surface is created during this blunting process, which is comparable in size to the ΔCTOD; with unloading there is reverse dislocation motion but the slip reversibility is never 100% and so some of this new surface is translated into crack extension.

Accordingly, mechanistically the crack growth per cycle should be proportional to the ΔCTOD, viz.:

$$\frac{da}{dN} = \beta\,\Delta\text{CTOD} \propto \beta\,\frac{\Delta K^2}{\sigma_y E}\,, \qquad (7.14)$$

where β is a rather ill-defined factor related to the efficiency of blunting and slip reversibility. Numerical studies for fatigue-crack growth driven by pure cyclic plasticity give $\beta \sim 0.3$ to 0.4 [26], which imply that 30%—40% of the new surface created by crack blunting can be translated into crack advance. The Paris slope of $m = 2$ in Eq. (7.14) is reasonable — most ductile metals display m exponents between 2 and 4 in the mid-range of growth rates (m is invariably equal to 2 if striation spacings are measured to represent da/dN) — and the inverse dependence of growth rates on the elastic modulus E is well known. However, the inverse dependence on yield strength is rarely seen; it has been reasoned that the yield strength also has an offsetting effect on β. Indeed, apart from the role of the modulus, strength level and microstructure have a remarkably little effect on striation fatigue-crack propagation in the mid-range of growth rates. Similarly, in contrast to regimes A and C, the mean stress, or R-ratio, has a relatively small effect on growth rates.

Near-threshold regime: In the near-threshold region A, where growth rates are typically less than $\sim 10^{-9}$ m/cycle, crack extension along the crack front tends to be highly non-uniform. Threshold ΔK_{TH} values are typically in the range of 2—10 MPa$\sqrt{m}$ for most metallic materials but, along with near-threshold growth rates, are markedly dependent on the R-ratio; they also tend to be dependent upon material properties such as the strength and microstructural features such as the grain size [e.g.,15]. In contrast to behavior in the mid-range of growth rates, this largely results from the predominance of a form of crack-tip shielding based on crack wedging between the crack faces — *crack closure* — due to crack surface oxidation and more generally the roughness of the crack surface. These phenomena tend to be promoted at near-threshold levels because the crack opening displacements are smaller at the low ΔK levels. This is described in greater detail in the following section.

7.4.3 Crack closure

The crack closure phenomenon, first described by Elber [27], involves the crack surfaces coming into contact during the unloading cycle before the minimum load is reached. Elber's initial experiment was to elastically load a specimen containing a fatigue crack where he noticed that the compliance changed above a certain load P_{cl}; above P_{cl}, the compliance correctly indicated the crack length but the crack appeared to be closed below P_{cl} as the crack surfaces came into contact (Fig. 7.10). He reasoned that although the global driving force was $\Delta K = K_{max} - K_{min}$, the effective stress-intensity range experienced at the crack tip was reduced to $\Delta K_{eff} = K_{max} - K_{cl}$, when K_{cl}, the closure stress intensity corresponding to P_{cl}, exceeded K_{min}. However, where $K_{cl} < K_{min}$, $\Delta K_{eff} = \Delta K$, such that the role of crack closure in affecting the effective ΔK at the crack tip was minimized.

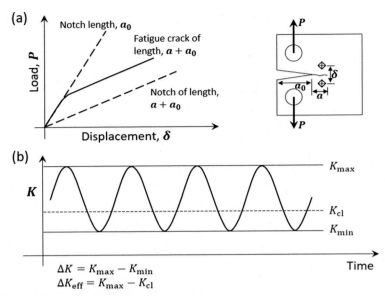

FIGURE 7.10 (a) Measurement of crack closure by detecting the closure load P_{cl} when the crack surfaces come into contact on unloading. (b) Corresponding definition of the effective stress-intensity range, ΔK_{eff}, actually experienced at the crack tip: when $K_{cl} > K_{min}$, $\Delta K_{eff} = K_{max} - K_{cl}$, where K_{cl} is the closure stress intensity corresponding to P_{cl}; when $K_{cl} < K_{min}$, $\Delta K_{eff} = \Delta K = K_{max} - K_{min}$.

Elber first identified the cause of crack closure to be associated with the residual stresses generated in the wake of the crack tip due to cyclic plasticity which causes the crack faces to prematurely come in contact [27]; this has been termed plasticity-induced crack closure. However, it subsequently became apparent that closure can result from several other wedging mechanisms inside the crack, which are especially important at near-threshold levels [28,29]. These are schematically illustrated in Fig. 7.11 along with the many other mechanisms of crack-tip shielding that can affect fatigue-crack propagation in different classes of materials, including metals, ceramics, and composites. Note that these mechanisms are all extrinsic (see also Section 6.4) in that they act at, or in the wake of, the crack tip to shield the crack from the far-field stresses (Fig. 6.6a).

For metal fatigue, the most important shielding process is crack closure, specifically from *plasticity-induced closure*, which is effective at the intermediate range of growth rates where the extent of plasticity is greater, but more importantly at near-threshold levels, from *oxide-induced closure* where corrosion products can form on the freshly exposed crack surfaces inside the crack, and from fracture surface *roughness-induced closure* which is the most pervasive mode of shielding in fatigue

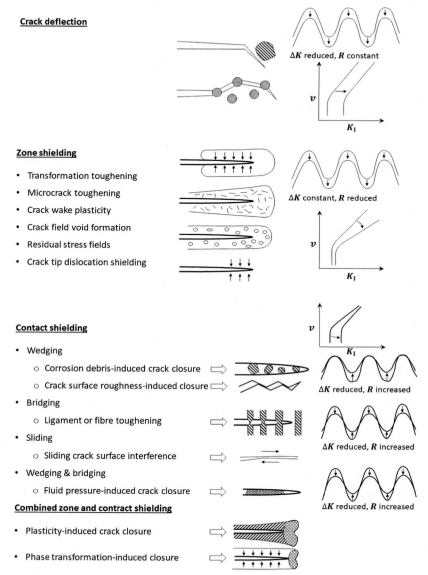

Crack deflection

Zone shielding

- Transformation toughening
- Microcrack toughening
- Crack wake plasticity
- Crack field void formation
- Residual stress fields
- Crack tip dislocation shielding

Contact shielding

- Wedging
    - ○ Corrosion debris-induced crack closure
    - ○ Crack surface roughness-induced closure
- Bridging
    - ○ Ligament or fibre toughening
- Sliding
    - ○ Sliding crack surface interference
- Wedging & bridging
    - ○ Fluid pressure-induced crack closure

Combined zone and contract shielding

- Plasticity-induced crack closure
- Phase transformation-induced closure

ΔK reduced, R constant

ΔK constant, R reduced

ΔK reduced, R increased

ΔK reduced, R increased

ΔK reduced, R increased

FIGURE 7.11 Schematic illustrations of the extrinsic mechanisms of crack-tip shielding that can enhance resistance to fatigue-crack growth, with indications of how each class of mechanisms — crack deflection, zone shielding, and contact shielding — can affect the v-K (da/dN vs. ΔK) curves. The most important mechanisms for fatigue-crack growth in metals are crack deflection and the crack closure mechanisms, principally oxide- and roughness-induced closure which are pure wedging mechanisms that are most effective at small CTODs, i.e., at low ΔKs, and plasticity-induced closure which is caused by cyclic plasticity which is more prevalent at higher ΔKs [29]. *Reproduced with permission.*

associated with wedging by rough fracture surface asperities. Note that the latter roughness-induced closure mechanism is promoted by crack deflection, which is also a shielding mechanism in its own right.

Fig. 7.11 also indicates how the presence of these shielding mechanisms can affect the v-K (da/dN vs. ΔK) curve. Crack deflection causes a multiplicative reduction in both K_{max} and K_{min} and so can be effective at all growth rates. Whereas plasticity-induced closure is most prevalent at intermediate to higher growth rates where there is more cyclic plasticity, oxide- and roughness-induced closure, conversely, are most effective at near-threshold levels where the crack opening displacements are small and comparable to the thickness of oxide films and/or fracture surface roughness.

Although the value of K_{cl}, and hence the precise ΔK_{eff} value, in the presence of crack closure is not easy to measure, and thus is not a design parameter, the presence of crack closure and the other forms of crack-tip shielding relevant to fatigue (Fig. 7.11) provide us with feasible explanations of many of the behavioral characteristics of fatigue cracks in the different regimes:

- *Effect of R-ratio (or K_{max})*: Crack closure provides a ready explanation why raising the mean stress or R-ratio can increase da/dN at a given ΔK. This is because a higher R-ratio will raise both K_{min} and K_{max}, meaning that the crack will be open for more of the cycle; once K_{min} exceeds K_{cl} the closure effect is nullified, and the crack will experience the full driving force as $\Delta K_{eff} \rightarrow \Delta K$. Such R-ratio effects are particularly significant at near-threshold levels (Fig. 7.12) because, as noted above, this is where the oxide- and roughness-induced mechanisms are most potent.
- *Microstructural effects of grain size*: Growth rates in the mid-growth rate regime, where the mechanism of crack advance is by striations (Section 7.4.2), are relatively insensitive to microstructural effects, as can be seen by the data on steels in Fig. 7.5. But like R-ratio effects, this is definitely not the case in the near-threshold regime, where growth rates can be markedly influenced by oxide- and especially roughness-induced crack closure. Coarse-grained structures generally display lower near-threshold growth rates and higher ΔK_{TH} thresholds. Why? ... because coarser structures tend to promote crack deflection and the more tortuous crack paths create rougher fracture surfaces to enhance closure. This is especially true in planar-slip materials, such as Al-Li alloys, where the crack path can be crystallographic and facetted, and in duplex microstructures, such as dual-phase steels [31], where a hard phase can act to deflect the crack path (Fig. 7.13). In fact, the crack-tip shielding for the particular alloy microstructures in Fig. 7.13 is especially effective

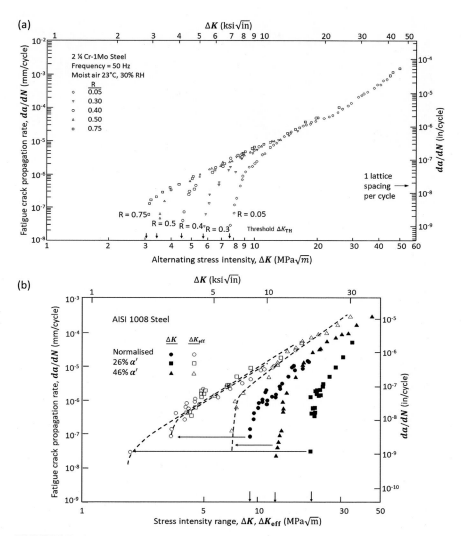

FIGURE 7.12 Variation of fatigue-crack growth rates, da/dN, as a function of the stress-intensity range, ΔK, for (a) bainitic 2¼Cr-1Mo pressure vessel steel SA542-3 in room air showing the marked effect of R-ratio (varying from 0.05 to 0.75) on near-threshold behavior, as compared to a minimal effect at the mid-range of growth rates [30], and (b) dual-phase ferritic-martensitic (α-α') AISI 1008 steel, tested at $R = 0.05$, showing the marked effect of crack closure for three microstructures in terms of the difference in the growth-rates data plotted as a function of $\Delta K_{eff} = K_{max} - K_{cl}$, as compared to the globally applied ΔK [31]. The *horizontal arrows* in (b) indicated the difference in ΔK_{eff} compared to ΔK at the threshold; as such they indicate the magnitude of the shielding effect from crack closure. *Reproduced with permission.*

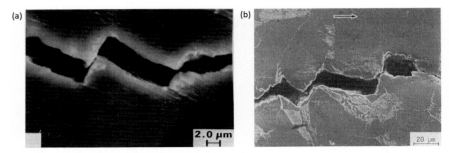

FIGURE 7.13 Scanning electron micrographs of the crack-path morphology during fatigue-crack growth at $R = 0.1$ in (a) a planar-slip Al-Li alloy 2090-T8 [29], and (b) a dual-phase AISI 1008 with a coarse, duplex ferritic-martensitic structure [31]. In both materials, the rough, deflected crack paths promote contact and wedging between the mating crack surfaces which induces significant crack closure. The *arrow* in (b) indicates the general direction of crack advance. *Reproduced with permission.*

with the magnitude of the ΔK_{eff} actually experienced at the crack tip being a factor of two or more smaller than the globally applied ΔK (Fig. 12b). These microstructural effects on ΔK_{TH} and near-threshold growth rates, however, are markedly diminished at high R-ratios; shielding by crack deflection remains but the larger crack opening displacements at the higher K_{max} values serve to compromise any crack surface contact.

- *Retardation due to variable amplitude loading*: Fatigue-crack growth subjected to a single tensile overload cycle (with the overload ΔK_{OL} typically at least twice that of the baseline ΔK) will experience an immediate increase in growth rate due to the high ΔK_{OL}, but this will be followed by a period of retardation, and sometimes even arrest, as the crack grows into the overload plastic zone (Fig. 7.14). Recall that unloading leaves a compressive zone at the crack tip (Fig. 7.6); the overload cyclic zone here will naturally be larger than the baseline zone and will lead to the retardation while the crack is propagating through it. Mechanistically, compressive residual stresses in the crack wake and the resulting plasticity-induced crack closure [32,33], coupled in certain materials by crack deflection, are the shielding mechanisms primarily responsible for the transient retardation in growth rates within the overload zone.

- *Small cracks*: The growth rates of small cracks, typically less than 1 mm, can often exceed those of large cracks at the same applied ΔK levels; in addition, small cracks can propagate at ΔK levels less than the ΔK_{TH} fatigue threshold measured for regular long cracks [28]. This can be clearly seen in the da/dN *vs.* ΔK plot for a Ti-6Al-4V alloy (Fig. 7.15), where the growth rates at $R = 0.1$ of small cracks in the

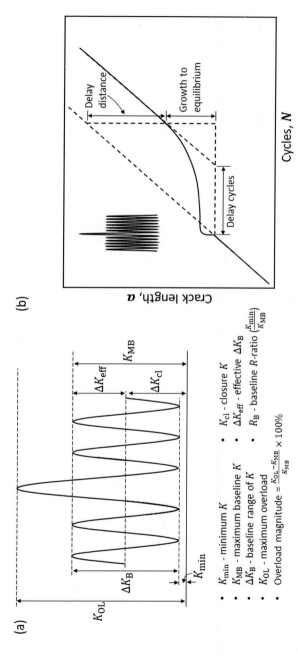

FIGURE 7.14 (a) Definition of parameters and (b) schematic of the crack length a vs. number of cycles N, under constant baseline ΔK conditions, for a single tensile overload illustrating the subsequent period of retardation defined in terms of the "delay cycles" and "delay distance."

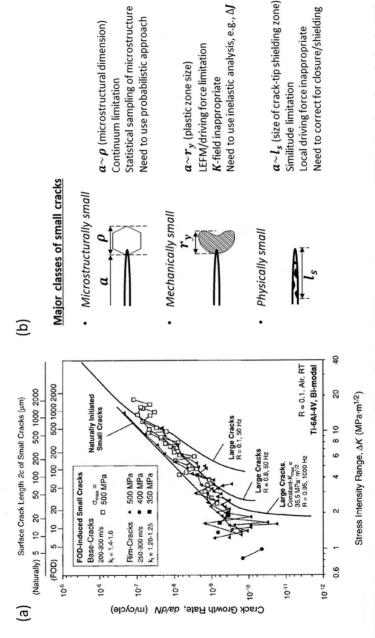

Major classes of small cracks

- *Microstructurally small*

  $a \sim \rho$ (microstructural dimension)
  Continuum limitation
  Statistical sampling of microstructure
  Need to use probabilistic approach

- *Mechanically small*

  $a \sim r_y$ (plastic zone size)
  LEFM/driving force limitation
  K-field inappropriate
  Need to use inelastic analysis, e.g., ΔJ

- *Physically small*

  $a \sim l_s$ (size of crack-tip shielding zone)
  Similitude limitation
  Local driving force inappropriate
  Need to correct for closure/shielding

FIGURE 7.15 (a) Fatigue-crack growth rates da/dN as a function of ΔK for naturally occurring and FOD-induced small cracks ($a \sim 5-1000$ μm) at $R = 0.1$, as compared to through-thickness large (>5 mm) cracks at $R = 0.1$, 0.8, and 0.95, in a bimodal Ti-6Al-4V alloy, showing small crack growth at ΔK values below the large-crack ΔK_{TH} thresholds [34]. (b) Schematic illustrations and characteristics [35] of the three major types of small cracks. Note in (a) that k_t is the effective stress concentration factor for the damage sites, e.g., FOD indentations, where the small cracks are initiated. *Reproduced with permission.*

range of sizes from 5 to 1000 μm (*data points*), both naturally occurring and induced by foreign-object damage (FOD), are compared with corresponding through-thickness large crack (>5 mm) data measured at R-values of 0.1, 0.8, and 0.95 (*lines*).

So what exactly are small cracks and why do they behave differently? The major types of small cracks are schematically illustrated in Fig. 7.15b, together with some of their characteristics. *Microstructurally-small cracks* are comparable in size with microstructural length-scales, e.g., the grain size. Since this is a dimension below the homogeneous continuum, fracture mechanics characterizations, e.g., using K, etc., are really not applicable, and statistical or probabilistic approaches need to be employed. *Mechanically-small cracks* are comparable with the plastic-zone size. This is a LEFM limitation, i.e., the K-field characterization of the crack-tip stresses and strains is invalid, and nonlinear-elastic approaches need to be considered. One such possibility is the use of ΔJ, and despite the uncertainty of using this parameter where the crack is continually subjected to unloading, some success has been achieved in characterizing small crack data measured under large-scale yielding conditions [19]. *Physically-small cracks* are of a size comparable with the steady-state length of the crack-tip shielding zone in the crack wake. In simple terms, at the same globally-applied ΔK level, the physically-small fatigue crack cannot develop as much crack closure in its wake (because its wake is limited). This is the most common reason for the small-crack effect. Indeed, although K_{cl} measurements are difficult, especially for small cracks, there has been some success in normalizing da/dN data for small and long cracks using ΔK_{eff}.

7.4.4 Measurement of fatigue-crack growth rates

ASTM Standard E647 [36] provides guidance for the measurement of fatigue-crack growth rates, using geometries that are similar to the plane-strain K_{Ic} Standard (Section 3.3), namely the compact-tension and middle-cracked tension samples, only without the stringent validity criteria. However, the fatigue test needs to be nominally elastic to justify a K-based characterization of growth rates, and so the one size requirement is that the remaining uncracked ligament b should satisfy the requirement:

$$b \geq \frac{4}{\pi}\left(\frac{K_{max}}{\sigma_y}\right)^2, \qquad (7.15)$$

where σ_y is the yield stress. However, σ_y can be replaced by the flow stress, $\sigma_o = \frac{1}{2}(\sigma_y + \sigma_{UTS})$, for high strain hardening materials. There is no size requirement on thickness.

There is a need of course to continuously monitor crack lengths throughout the test. This can be done optically, using elastic unloading compliance measurements using a back-face strain or mouth-opening displacement gauges, or by using the electrical-potential technique[2] which involves passing a constant current through the specimen and measuring the change in electrical potential across the notch.

Fatigue-crack growth testing is generally performed in load control at a constant R-ratio, under conditions of (i) increasing ΔK — this is generally a constant load test — or (ii) decreasing ΔK, which involves programmed load shedding. Constant load testing is perfectly fine for measuring intermediate to high growth rates, but it is almost impossible to determine the fatigue threshold this way. Accordingly, to measure near-threshold growth rates and the ΔK_{TH} value, a programmed decrease in load is used to slowly reduce the ΔK level until growth rates are less than $\sim 10^{-10}$–10^{-11} m/cycle (which defines ΔK_{TH}). To avoid premature arrest by reducing the load too fast, as the prior (higher) K level would have a larger cyclic plastic zone, ASTM prescribes a specific normalized K-gradient, $\frac{1}{K}\left(\frac{dK}{da}\right)$, for the decrease in K which must exceed -0.08 mm^{-1}.

Thresholds at very high R-ratios above ~ 0.9 can be measured by load shedding under conditions of constant K_{max}/increasing K_{min} (the $R = 0.95$ data in Fig. 7.15a was collected in this way). The measured near-threshold growth rates will be under variable R-conditions, but this is the optimal way to define ΔK_{TH} thresholds above $R \sim 0.9$.

To determine the growth rates, you need to find the gradient of the a vs. N curves at specific points. The ASTM Standard suggests two numerical methods to do this involving (i) simply finding the slope between two consecutive points or (ii) using an incremental-step (overlapping) polynomial method, where you fit a quadratic to, say, eight points, then differentiate to determine da/dN for the middle two points, move forward two points and repeat the process. Full details are given in the Standard.

7.4.5 Fatigue-crack growth in brittle materials

So far we have talked about the fatigue of ductile materials such as metallic structures. But what about brittle materials like ceramics? Do they fatigue? The short answer to that question is that brittle materials are far less susceptible to fatigue. For example, prototypical brittle materials,

---

[2] The electrical potential method [37], using DC or AC currents, is a particularly effective technique but can suffer from electrical shorting across the crack, especially in low R-ratio tests, and so it is prudent to check the method, indeed any crack monitoring method, with periodic optical measurements of the crack size; this can also identify whether crack tunneling is taking place.

such as diamond, silicon[3] and soda lime glass, simply fracture with little to no evidence of preceding cyclic damage. However, brittle materials that are toughened by the extrinsic mechanisms listed in Fig. 6.7 can become susceptible, as fatigue can act to degrade this toughening under cyclic loads [39].

Fig. 7.16a shows a comparison of the fatigue-crack growth behavior in metals (high-strength steels, aluminum and titanium alloys) as compared to intermetallics (α_2-Ti$_3$Al, γ-TiAl, MoSi$_2$) and ceramics (Al$_2$O$_3$, Mg-PSZ, pyrolytic carbon). Moving from right to left in this graph with materials of increasing brittleness in terms of a lower toughness, the trend is for the fatigue threshold to be decreased and the slope of the da/dN vs. ΔK curve, i.e., the Paris exponent, to increase; it is typically m ~2 to 4 for metals (in the mid-range of growth rates), ~10 to 20 for intermetallics, and ~15 to 50 or more for ceramics. Such high exponents have broad implications for the life prediction of structural components manufactured from these materials as high m values can make the projected safe life overly sensitive to stress and defect sizes, as discussed in Section 7.5.

Why is there such a difference though? In brittle materials, it is clear that with rare exception, the mechanism of crack advance under cyclic loading in fatigue is identical to that for non-cyclic overload fracture. The difference, as mentioned above, is that cyclic loading can induce a progressive degradation in the toughening, i.e., the shielding mechanisms behind the crack tip, which acts to locally elevate the near-tip driving force. For example, Al$_2$O$_3$ and Si$_3$N$_4$ ceramics can be significantly toughened by interlocking grain bridging (Section 6.4.2) but the continued opening and closing of the crack in fatigue can lead to wear of the grain interfaces which progressively diminishes the potency of the bridging effect [40]; a similar phenomenon can be seen in fiber-toughened ceramics, such as the silicon carbide composites reinforced with SiC fibers, which are currently replacing certain metallic parts in gas-turbine engines. As shown schematically in Fig. 7.16b, it is this cyclic suppression of extrinsic crack-tip shielding that is the principal source for the susceptibility of brittle materials to cyclic fatigue failure. By contrast, the propagation of fatigue cracks in metallic materials involves primarily intrinsic damage processes occurring ahead of the crack tip, i.e., involving striation crack advance by crack-tip blunting and resharpening (Section 7.4.2), which is clearly a mechanism distinct from fracture under monotonic loads. Shielding mechanisms, principally in

---

[3] Silicon in bulk form does not fatigue, but with micron-scale silicon, used in MEMS devices, the ubiquitous SiO$_2$ oxide film is no longer an insignificant part of the structure. As SiO$_2$ is susceptible to stress-corrosion cracking in moist air, under very high-cycle conditions premature fatigue failure can occur at stresses on the order of one half of silicon's monotonic fracture strength [38].

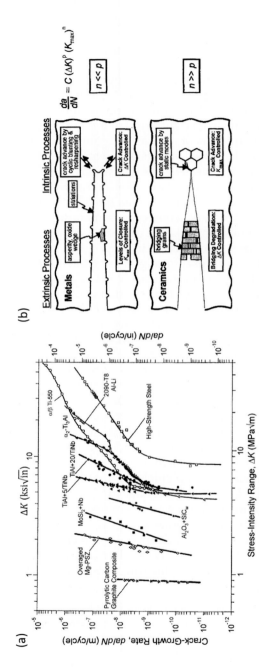

FIGURE 7.16 Fatigue-crack propagation in ductile and brittle solids, showing (a) variation in crack-growth rates, da/dN, as a function of ΔK for metals, intermetallics and ceramics, and (b) intrinsic and extrinsic mechanisms involved in fatigue-crack growth in ductile (metals) and brittle (ceramics), showing the relative dependencies of growth rates on ΔK and K_{max} [39]. *Reproduced with permission.*

the form of crack closure, can however act in the crack wake and are limited at increasing K_{max}.

This distinction in the fatigue processes in ductile and brittle materials has implications for the form of the crack-growth law. As noted, in terms of the simple Paris law, $da/dN = C \Delta K^m$, the exponent m is 2 to 4 for metals yet far higher for ceramics, which we can explain by rewriting Eq. (7.1) in terms of both K_{max} and ΔK [39]:

$$\frac{da}{dN} = C'' \Delta K^p K_{max}^n , \qquad (7.16)$$

where in terms of the scaling constants in Eq. (7.1), $n + p = m$ and $C'' = C(1 - R)^n$. The dominant term in this expression pertains to the crack advance mechanism.

Since the physical mechanisms of crack advance and crack-tip shielding are quite different in metals and ceramics, the dependencies on the alternating and mean loads, specifically ΔK and K_{max}, are also quite different. Clearly, in metals the dominant dependence of ΔK is a consequence of the intrinsic crack-advance mechanism driven by cyclic plasticity; the smaller K_{max} dependence results primarily from its effect on crack-opening displacements, which in turn controls the degree of crack wedging due to crack closure in the wake. Thus, *for ductile metals, $p \gg n$* in Eq. (7.16). An example for this is a Ni-base superalloy where $\frac{da}{dN} \propto \Delta K^3 K_{max}^{0.4}$.

In ceramics, conversely, growth rates are principally a function of K_{max}, since the crack-advance mechanism is identical to that under static loading; the much weaker ΔK dependence here arises from the cyclic-induced degradation in shielding in the wake. Thus, *for ceramics, $p \ll n$.* An example, for a monolithic silicon nitride, is $\frac{da}{dN} \propto \Delta K^{1.3} K_{max}^{29}$.

In intermetallics, fatigue properties are intermediate between these two extremes, although these materials span the range from being reasonably ductile, like α_2-Ti$_3$Al, where $p > n$, to being distinctly brittle, like MoSi$_2$, where $p < n$; in general though for intermetallics, n is more likely to be comparable to p. An example, for a Nb-wire reinforced MoSi$_2$ is $\frac{da}{dN} \propto \Delta K^8 K_{max}^{13}$.

Thus, although much progress has been made over the last few decades in significantly improving the fracture resistance of ceramics and intermetallics using extrinsic toughening mechanisms, unfortunately the apparent ductility that this induces does make these materials prone to fatigue.

7.4.6 Corrosion fatigue-crack growth

One of the most virulent forms of subcritical cracking is corrosion fatigue associated with the conjoint action of environmentally-assisted cracking and fatigue-crack growth. The many mechanisms of corrosion fatigue, whether involving aqueous corrosion, hydrogen-assisted cracking, or higher temperature oxidation and/or sulfidation, are too numerous to describe here [5,41], but from a fracture mechanics perspective they can generally be considered in terms of a synergistic or additive effect of a largely cycle-dependent mechanical fatigue process with a time-dependent environmental process. In this context, McEvily and Wei [41] have classified these processes as (cycle-dependent) true-corrosion fatigue, (time-dependent) stress-corrosion fatigue, and (cycle- and time-dependent) corrosion-fatigue (Fig. 7.17).

True corrosion fatigue (Fig. 7.17a) is essentially a cycle-dependent acceleration of the mechanical fatigue process due to environmental factors, by such phenomena as the hydrodynamic action of the opening and closing of the crack to drive a corrosive fluid to the crack tip, the cycle-dependent creation of fresh surface at the crack tip for hydrogen adsorption or an oxidation reaction to occur, and the oxidation of these surfaces to limit slip-reversibility at the crack tip. One simple way to represent this is to consider the corrosion fatigue-crack growth rate $(da/dN)_{cf}$ as the mechanical fatigue-crack growth rate in an inert environment $(da/dN)_{fat}$ multiplied by an acceleration factor, Ω:

$$\left(\frac{da}{dN}\right)_{cf} = \Omega\left(\frac{da}{dN}\right)_{fat}. \tag{7.17}$$

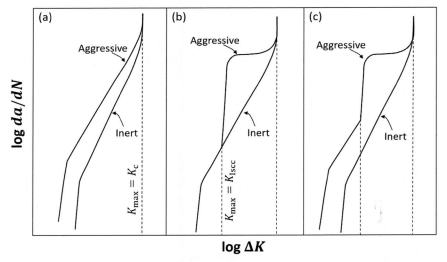

FIGURE 7.17   Basic types of corrosion fatigue crack growth behavior: (a) cycle-dependent true-corrosion fatigue, (b) time-dependent stress-corrosion fatigue, and (c) cycle- and time-dependent corrosion fatigue.

Time-dependent stress-corrosion fatigue (Fig. 7.17b), on the other hand, can be modeled as a direct superposition of the time-dependent (static) stress-corrosion (or hydrogen-assisted) cracking mechanism [42], as depicted by a v-K curve, as in Figs. 7.2 and 7.3 with a crack-growth relationship of the form $(da/dt)_{scc} = C'K^{n'}$, onto the cycle-dependent fatigue process, represented by a Paris law of the form $(da/dN)_{fat} = C\,\Delta K^{m}$, when K_{max} exceeds the environmental threshold K_{Iscc}:

$$\left(\frac{da}{dN}\right)_{cf} = \left(\frac{da}{dN}\right)_{fat} + \frac{1}{\vartheta}\left(\frac{da}{dt}\right)_{scc}, \qquad (7.18)$$

where ϑ is the cyclic frequency.

An example of both types of behavior can be seen in Fig. 7.18a for fatigue crack growth in a 2¼Cr-1Mo pressure vessel steel cycled in air and gaseous hydrogen. The near-threshold regime below roughly 10^{-6} mm/cycle represents an example of true-corrosion fatigue, whereas above a $K_{max} \sim 20$ MPa$\sqrt{m}$, the static v-K curve for hydrogen-assisted cracking under non-cyclic loads can simply be superimposed on the mechanical fatigue curve.

Note that in the superimposition of a da/dt vs. K curve on a da/dN vs. ΔK fatigue curve, the location of the superimposed da/dt vs. K portion will depend on the frequency ϑ, as $(da/dt) \equiv \vartheta(da/dN)$, and the R-ratio, as $K_{max} = \Delta K/(1 - R)$. This can be seen in Fig. 7.18b, which shows the same steel fatigue tested in air and hydrogen gas but at different R-ratios from 0.05 to 0.5; the superposition of the static da/dt vs. K curve onto the fatigue data can now be seen to be occurring at different ΔK levels.

The superposition model presented here relies on the presumption that the mechanical and environmental mechanisms are simultaneous and essentially independent, such that the faster process dominates. However, one can envisage such a combination of mechanical and environmental mechanisms that are sequential and hence inter-dependent; in this case, the slower process will dominate. Under these conditions, a process competition model [44], rather than a superposition model, is better suited to describe stress-corrosion fatigue, with the mechanical and environmental contributions added in parallel, rather than in series:

$$\left(\frac{da}{dN}\right)_{cf}^{-1} = \left(\frac{da}{dN}\right)_{fat}^{-1} + \left(\frac{1}{\vartheta}\left(\frac{da}{dt}\right)_{scc}\right)^{-1}. \qquad (7.19)$$

7.5 Damage-tolerant life prediction

Until the 1970s, life prediction methodologies for loaded structures relied primarily on a stress- (or strain-) based total life approaches, as, for

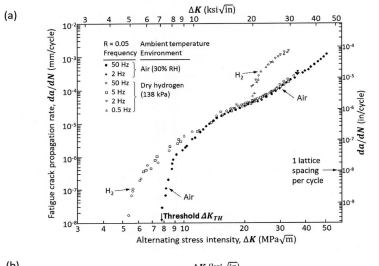

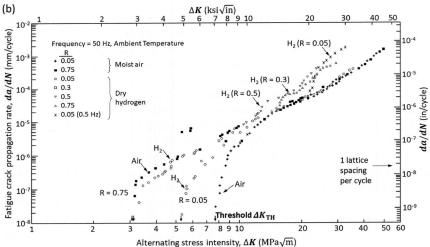

FIGURE 7.18   Fatigue-crack propagation rates, *da/dN*, as a function of ΔK for a bainitic 2¼Cr-1Mo pressure vessel steel SA542-3 tested in air and hydrogen gas at $R = 0.05$, showing (a) true-corrosion fatigue below ~10^{-6} mm/cycle, and stress-corrosion fatigue represented by the superimposition of the static *da/dt vs. K* curve in hydrogen above a $K_{max} \sim 20$ MPa√m; (b) the latter behavior for fatigue-crack growth at variable *R*-ratios from 0.05 to 0.5 [43]. *From S. Suresh, R.O. Ritchie. Mechanistic dissimilarities between environmentally influenced fatigue crack propagation at near-threshold and higher growth rates in lower strength steels, Metal Sci. 16 (1982) 529; copyright ©Institute of Materials, Minerals and Mining, reprinted by permission of Informa UK Limited, trading as Taylor & Francis Group, www.tandfonline.com on behalf of Institute of Materials, Minerals and Mining.*

example, with the use of *S-N* curves for fatigue-life prediction involving the initiation and propagation of cracks to failure in smooth-bar specimens. However, uncertainties whether such small laboratory test samples could truly predict the likelihood of crack initiation from a defect in a much larger structure led to the adoption, by U.S. Air Force and subsequently numerous other industries, of a fracture-mechanics based damage-tolerant life approach. Here the inevitable presence of defects in any structure is assumed; life is then based solely on the propagation of the largest undetected crack to failure, as, for example, defined by the limit load or when $K_I = K_c$. Hence, the presence of incipient cracks in structures is tolerated — that is of course where the name "damage-tolerant design" comes from; life prediction is thus a question of how large these cracks are and how fast they will grow until the structure fails, i.e., dependent on the rate of subcritical crack growth.

In this chapter, we have described the primary mechanisms of subcritical cracking, namely, by fatigue, creep, and environmentally-assisted cracking, although in practice these mechanisms are often combined, e.g., as creep-fatigue, corrosion fatigue, etc. We have also described the various crack-growth "laws," where the crack velocities, either with respect to time or number of cycles, are correlated to the relevant characterizing parameter governing their crack-tip stress and strain fields, e.g., involving relationships of the form of Eq. (7.1) for fatigue, Eq. (7.2) for environmentally-assisted cracking and Eq. (7.7) for creep. These relationships, or perhaps more complicated versions of them, then form the basis of the damage-tolerant life-prediction methodology for many safety-critical structures.

The way that damage-tolerant life-prediction procedures work is by first identifying critical locations on your structure; these could be locations where your stress analysis has indicated that there are regions of high stress, where there are notches that could act as stress concentrators, or where you have seen evidence of cracking in service. For these locations, you would then consider a worst-case defect, and determine the stress intensity solution for this crack configuration in light of the applied stresses that this part may experience in service — this is the first input to the life prediction analysis. We will assume here for clarity a simple mode I K_I-solution, in terms of the applied stress σ^∞ and crack size a:

$$K_I = Y\sigma^\infty \sqrt{\pi a}, \tag{7.20}$$

where Y is the geometry factor but recognize that for real structure, the K-solution could be far more complex, as shown in Box 3.2.

We next need to know how fast cracks will grow under these conditions in service. Let us assume that we have identified fatigue as the principal source of subcritical cracking; accordingly, we would need to determine the v-K curve for the material in question under conditions of stress state, R-ratio, environment, frequency, etc., that as closely match the service conditions as possible. For simplicity, we will assume here a Paris law growth-rate relationship:

$$\frac{da}{dN} = C\Delta K^m ,\tag{7.21}$$

again recognizing that the actual relationship for the service conditions could be far more complex. Substituting the K_I-solution (Eq. 7.20) into the crack-growth relationship Eq. 7.21 gives:

$$\frac{da}{dN} = C(Y\Delta\sigma\sqrt{\pi a})^m ,\tag{7.22}$$

where $\Delta\sigma$ is the *in-service* alternating stress. Rearranging and integrating from the initial crack size, a_i, to a final crack size, a_f, gives:

$$\int_{a_i}^{a_f} \frac{da}{a^{\frac{m}{2}}} = CY^m\pi^{\frac{m}{2}}\Delta\sigma^m\int_0^{N_f} dN ,\tag{7.23}$$

such that the number of cycles to failure, N_f, is given by (for $m \neq 2$):

$$N_f = \frac{2}{(m-2)Y^mC\pi^{\frac{m}{2}}\Delta\sigma^m} \left[\frac{1}{a_i^{(m-2)/2}} - \frac{1}{a_f^{(m-2)/2}} \right] .\tag{7.24}$$

The projected lifetime can be seen to be a strong function of the stress, which can be raised to a power of 2–4, or higher in brittle materials, and of the initial crack size, a_i, which is generally related to the estimated size of the largest undetected crack. Quantifying this largest undetected crack size generally mandates the use of non-destructive evaluation (NDE) techniques, such as fluorescent dye penetrants, ultrasonic testing, X-radiography, eddy-current testing, etc., prior to the component entering service; the value of a_i is then equated to the limit of resolution of the technique that is used (ideally, to the largest crack that can be missed!). Such NDE can then be carried out periodically on critical parts, e.g., turbine disks and blades in aircraft engines, periodically throughout the life of the component, to verify the lifetime predictions and to make repairs if deemed necessary.

The final crack size, a_f, can be equated to the crack size where the stress in the remaining ligament exceeds the tensile strength, i.e., at the limit load, or at final fracture when $K_{max} = K_c$, i.e., where a_f exceeds the critical crack size given by $a_c = \frac{1}{\pi}\left(\frac{K_c}{Y\sigma_{max}}\right)^2$. The value of a_f generally has a marginal effect on the life, as it is obviously larger than a_i and in a reciprocal term; however, its magnitude is critical from the perspective of whether it can be detected prior to the structure fracturing (see also Section 3.3.8).

This damage-tolerant life-prediction strategy is designed to give a conservative estimate of the life, by considering worse-case conditions and assuming that the crack initiation life is zero. However, the marked sensitivity of fatigue-crack growth rates to the applied stress intensity in brittle materials, both at elevated and especially ambient temperatures, presents unique challenges for damage-tolerant design and life-prediction methods for structural components fabricated from these materials [39]. For advanced materials such as intermetallics and ceramics subject to cyclically varying loads, this approach may be difficult, if not impossible, to implement in practice due to the large values of the exponent m in their "crack-growth laws." Since the projected life in Eq. (7.24) is proportional to the reciprocal of the applied stress raised to the power of m, a factor of two change in applied stress can lead to projections of the life of a ceramic component (where m can be as high as 20 or more) to vary by more than six orders of magnitude. Essentially, because of the high exponents, the life spent in crack propagation in advanced materials will either be extremely limited or infinitely large, depending upon whether the initial stress intensity is just above or below the threshold stress intensity.

Accordingly, a more appropriate approach for these materials may be to design on the basis of a threshold below which failure cannot occur. For fatigue, this may involve a conventional endurance strength, or fatigue limit, determined from S-N data or, more conservatively, the ΔK_{TH} or $K_{max,TH}$ thresholds for no crack growth. Indeed, the latter use of a threshold stress intensity in design to define a critical crack size, rather than at fracture instability, is not unusual, particularly where subcritical growth rates are extremely fast, or for fatigue at very high frequencies, as once above the threshold a crack can grow along the entire v-K curve to failure in an unacceptably short time (e.g., this can be the case for turbine blades subjected to vibration fatigue).

However, even these approaches may not be completely conservative due to uncertainties in the definition of such thresholds in the presence of small cracks [45]. The question of the small crack effect in these instances is essential to consider, either by making the threshold measurements on relatively small cracks (the engine companies often using small tensile-type specimens containing small "cord" starter cracks to measure near-threshold data in the design of their turbine blades), and/or by using crack growth experiments where the role of crack closure can be limited. This can be achieved by using ΔK_{eff} measurements, but more simply by using very high R-ratio crack growth data. Indeed, the large-crack, constant K_{max} growth rates, at up to $R = 0.95$ in Fig. 7.15a, do provide a closer estimate, albeit still approximate, to the actual small crack results.

Notwithstanding some of these difficult issues associated with brittle materials and small cracks, the fracture mechanics-based damage-tolerant life-prediction methodology is designed to give conservative estimates of the safe lifetimes of structures and components even where periodic NDE inspections are not feasible, e.g., for medical prostheses. Consequently, it is now widely used in the nuclear and commercial/military aerospace industries and is increasingly finding application for the design and life prediction of numerous components subjected to monotonic or cyclically varying loads in extreme temperature and/or aggressive environments.

References

[1] P.C. Paris, M.P. Gomez, W.F. Anderson, A rational analytic theory of fatigue, Trend Eng. 13 (1961) 9.

[2] P.C. Paris, F. Erdogan, A critical analysis of crack propagation laws, J. Basic Eng. 85 (1963) 528.

[3] H.H. Johnson, P.C. Paris, Sub-critical flaw growth, Eng. Fract. Mech. 1 (1968) 3.

[4] H. Hänninen, Stress corrosion cracking, in: I. Milne, R.O. Ritchie, B. Karihaloo (Eds.), Comprehensive Structural Integrity, vol. 6, Elsevier, Oxford, UK, 2003, p. 1.

[5] R.P. Gangloff, Hydrogen-assisted cracking in high-strength alloys, in: I. Milne, R.O. Ritchie, B. Karihaloo (Eds.), Comprehensive Structural Integrity, vol. 6, Elsevier, Oxford, UK, 2003, p. 31.

[6] R.O. Ritchie, M.H. Castro-Cedeno, V.F. Zackay, E.R. Parker, Effects on silicon additions and retained austenite on stress corrosion cracking in ultra-high strength steels, Metall. Trans. A 9A (1978) 35.

[7] J.D. Landes, J.A. Begley, A fracture mechanics approach to creep crack growth, in: ASTM STP 590, American Society for Testing and Materials, West Conshohocken, PA (1976) 128.

[8] K.N. Nikbin, G.A. Webster, C.E. Turner, Relevance of nonlinear fracture mechanics to creep crack growth, in: ASTM STP 601, American Society for Testing and Materials, West Conshohocken, PA, 1976, p. 47.

[9] A. Saxena, Advanced Fracture Mechanics and Structural Integrity, CRC Press, Boca Raton, FL, 2019.

[10] H. Riedel, J.R. Rice, Tensile cracks in creeping solids, in: ASTM STP 700, American Society for Testing and Materials, West Conshohocken, PA, 1980, p. 112.

[11] K. Ohji, K. Ogura, S. Kubo, Stress-strain fields and modified J-integral in the vicinity of the crack tip under transient creep conditions, Trans. Japan Soc. Mech. Eng. 790−13 (1979) 18.

[12] A. Saxena, Creep crack growth under nonsteady-state conditions, in: ASTM STP 905, American Society for Testing and Materials, West Conshohocken, PA, 1986, p. 185.

[13] J.L. Bassani, D.E. Hawk, A. Saxena, in: ASTM STP 995, in: Evaluation of the Ct Parameter for Characterizing Creep Crack Growth Rate in the Transient Regime, vol. I, American Society for Testing and Materials, West Conshohocken, PA, 1990, p. 112.

[14] A. Saxena, Nonlinear Fracture Mechanics for Engineers, CRC Press, Boca Raton, FL, 1997.

[15] R.O. Ritchie, Influence of microstructure on near-threshold fatigue crack propagation in ultra-high strength steel, Met. Sci. 11 (1977) 368.

[16] R.G. Forman, V.E. Kearney, R.M. Engle, Numerical analysis of crack propagation in cyclic-loaded structures, J. Basic Eng. 89 (1967) 459.

[17] M. Klesnil, P. Lukáš, Influence of strength and stress history on growth and stabilisation of fatigue cracks, Eng. Fract. Mech. 4 (1972) 77.

[18] R.J. Donahue, H.M. Clark, P. Atanmo, R. Kumble, A.J. McEvily, Crack opening displacement and the rate of fatigue crack growth, Int. J. Fract. Mech. 8 (1972) 209.

[19] N.E. Dowling, J.A. Begley, Fatigue crack growth during gross plasticity and the J-integral, in: ASTM STP 590, American Society for Testing and Materials, West Conshohocken, PA, 1976, p. 82.

[20] J.R. Rice, Mechanics of crack-tip deformation and extension by fatigue, in: ASTM STP 415, American Society for Testing and Materials, West Conshohocken, PA, 1967, p. 247.

[21] R.O. Ritchie, J.F. Knott, Mechanisms of fatigue crack growth in low alloy steel, Acta Metall. 21 (1973) 639.

[22] C. Laird, G.C. Smith, Crack propagation in high stress fatigue, Philos. Mag. A 7 (1962) 847.

[23] R.O. Ritchie, Near-threshold fatigue crack propagation in steels, Int. Met. Rev. 24 (1979) 205.

[24] R.M.N. Pelloux, Crack extension by alternating shear, Eng. Fract. Mech. 1 (1970) 697.

[25] P. Neumann, New experiments concerning the slip processes at propagating fatigue cracks — I, Acta Metall. 22 (1974) 1155.

[26] Z.S. Hosseini, M. Dadfarnia, B.P. Somerday, P. Sofronis, R.O. Ritchie, On the theoretical modeling of fatigue crack growth, J. Mech. Phys. Solid. 121 (2018) 341.

[27] W. Elber, Fatigue crack closure under cyclic tension, Eng. Fract. Mech. 2 (1970) 37.

[28] S. Suresh, R.O. Ritchie, The propagation of short fatigue cracks, Int. Met. Rev. 29 (1984) 445.

[29] R.O. Ritchie, Mechanisms of fatigue crack propagation in metals, ceramics and composites: role of crack-tip shielding, Mater. Sci. Eng. 103 (1988) 15.

[30] S. Suresh, R.O. Ritchie, On the influence of environment on the load ratio dependence of fatigue thresholds in pressure vessel steel, Eng. Fract. Mech. 18 (1983) 785.

[31] J.-K. Shang, J.-L. Tzou, R.O. Ritchie, Role of crack tip shielding in the initiation and growth of long and small fatigue cracks in composite microstructures, Metall. Trans. A. 18A (1987) 1613.

[32] E.F.J. von Euv, R.W. Hertzberg, R. Roberts, Delay effect in fatigue-crack propagation, in: ASTM STP 513, American Society for Testing and Materials, West Conshohocken, PA, 1972, p. 230.

[33] J.C. Newman, Prediction of fatigue crack growth under variable amplitude and spectrum loading using a closure model, in: ASTM STP 761, American Society for Testing and Materials, West Conshohocken, PA, 1982, p. 255.

[34] J.O. Peters, R.O. Ritchie, Influence of foreign-object damage on crack initiation and early crack growth during high-cycle fatigue of Ti-6Al-4V, Eng. Fract. Mech. 67 (2000) 193.

[35] R.O. Ritchie, J. Lankford, Small fatigue cracks: a statement of the problem and potential solutions, Mater. Sci. Eng. 84 (1986) 11.

[36] ASTM Standard E647-15, Standard method for measurement of fatigue crack growth rates, in: Annual Book of ASTM Standards, American Society for Testing and Materials, West Conshohocken, PA, 2020, 3.01.

[37] G.H. Aronson, R.O. Ritchie, Optimization of the electrical potential technique for crack growth monitoring in compact test pieces using finite element analysis, ASTM J. Test. Eval. 7 (1979) 208.

[38] C.L. Muhlstein, E.A. Stach, R.O. Ritchie, A reaction-layer mechanism for the delayed failure of micron-scale polycrystalline silicon structural films subjected to high-cycle fatigue loading, Acta Mater. 50 (2002) 3579.

[39] R.O. Ritchie, Mechanisms of fatigue-crack propagation in ductile and brittle solids, Int. J. Fract. 100 (1999) 55.

[40] R.H. Dauskardt, A frictional wear mechanism for fatigue-crack growth in grain bridging ceramics, Acta Metall. Mater. 41 (1993) 2765.

[41] A.J. McEvily, R.P. Wei, Fracture mechanics and corrosion fatigue, in: Corrosion Fatigue: Chemistry, Mechanics and Microstructures. Proc. NACE Conf., NACE, Houston, TX, 1972, p. 25.

[42] R.P. Wei, J.D. Landes, Correlation between sustained-load and fatigue crack growth in high strength steels, Mater. Res. Stand. 9 (1969) 25.

[43] S. Suresh, R.O. Ritchie, Mechanistic dissimilarities between environmentally-influenced fatigue crack propagation at near-threshold and higher growth rates in lower strength steels, Met. Sci. 16 (1982) 529.

[44] I.M. Austen, P. McIntyre, Corrosion fatigue of high-strength steel in low-pressure hydrogen gas, Met. Sci. 13 (1979) 420.

[45] R.O. Ritchie, J.O. Peters, Small fatigue cracks: mechanics, mechanisms and engineering applications, Mater. Trans. 42 (2001) 58.

8

Practical examples

8.1 Introduction

In this final chapter, we provide several worked examples of the use of fracture mechanics, drawn from our own practical experience. These range from deciding the choice of steel for a support rod to hold a large weight and the use of the leak-before-break concept for the design of a pressure vessel to the issue of the premature failure of pre-tensioned bolts and the estimation of the service life of medically implanted heart valve prostheses.

8.2 Worked examples

8.2.1 Support rod to carry a large weight

Our first example is an apparently simple one involving the choice of steel to make a support rod to carry a heavy weight. This could be the cable for a crane or in this particular case a rectangle rod to carry a weight P, as shown in Fig. 8.1a. For simplicity, let us make the cross-sectional dimensions of the rod to have a width $W = 2.1$ in. (53.3 mm) with a thickness B of 1 in. (25.4 mm). Our decision is whether to make it out of a low-strength mild steel, with a yield strength σ_y of 55 ksi (739 MPa) and a fracture toughness of $K_{Ic} = 100$ ksi$\sqrt{\text{in}}$ (110 MPa$\sqrt{\text{m}}$), or a high-strength low-alloy steel, with a much higher yield strength, σ_y of 250 ksi (1724 MPa), but with a fracture toughness of $K_{Ic} = 50$ ksi$\sqrt{\text{in}}$ (55 MPa$\sqrt{\text{m}}$).

One way to approach this problem is to calculate how much weight the rod can support without failure. However, failure can occur not only by fracture but also by plastic yielding (limit load failure) and so both scenarios need to be considered.

To first consider fracture, it would be prudent to believe that there will be defects in any piece of metal. In view of the seriousness of this

139

8. Practical examples

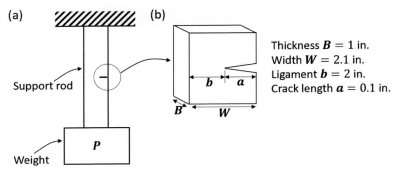

FIGURE 8.1 Schematic illustration of (a) a support rod carrying a weight P, which can be modeled as (b) a semi-infinite edge-cracked tension body, of width $W = 2.1$ in., thickness $B = 1$ in., containing a through-thickness edge crack of length $a = 0.1$ in.

application - let us say that people will be walking under the supported weight - we would need to carry out a nondestructive evaluation (NDE) of the finished part prior to putting it into service to look for any such defects. We will assume for simplicity here that the NDE technique that we used had a reliable limit of resolution of, say, 0.1 in. for a surface crack. If we take this as the largest undetected crack size to represent a worst-case through-thickness surface crack, we can model the support rod as an edge-cracked tension body, as in Fig. 8.1b, with thickness $B = 1$ in. (25.4 mm), width $W = 2.1$ in. (53.3 mm), crack length $a = 0.1$ in. (2.5 mm), and uncracked ligament $b = 2$ in. (50.8 mm). From the K-solutions in Fig. 3.3 in Box 3.2, the mode I K_I for this crack is given by:

$$K_I = 1.12 \, \sigma^\infty \sqrt{\pi a} , \tag{8.1}$$

where the far-field tensile stress in the structure is given by:

$$\sigma^\infty = P/bB . \tag{8.2}$$

For failure by fracture, we can consider $K_I = K_{Ic}$, such that the load at fracture, P_F, will be:

$$P_F = \frac{bB}{1.12} \frac{K_{Ic}}{\sqrt{\pi a}} , \tag{8.3}$$

and for failure by yielding, when $\sigma^\infty = \sigma_0$, the tensile (or yield) strength, the load at general yielding, P_y, will be approximately:

$$P_y = bB\sigma_y . \tag{8.4}$$

where $\sigma_0 = \sigma_y$, the yield strength, is taken to be conservative. Using Eqs. (8.3) and (8.4), the maximum weight that the support rod could carry before either fracture or general yielding are given in Table 8.1.

ok=sss.

....

TABLE 8.1 Mechanical properties and predicted failure loads, by yielding and fracture for the low-strength mild steel and high-strength low-alloy steel.

Steel	σ_y	K_{Ic}	P_y[a]	P_F
Low-strength steel	55 ksi (939 MPa)	100 ksi√in (110 MPa/m)	110,000 lb (489 kN)	320,000 lb (1423 kN)
High-strength steel	250 ksi (1724 MPa)	50 ksi√in (55 MPa√m)	500,000 lb (2224 kN)	160,000 lb (712 kN)

[a]to be conservative, we have assumed the yield strength in this calculation.

Examining these results clearly shows that the high-strength steel can support higher loads; it is predicted to fracture at loads exceeding 160,000 lb (712 kN), whereas the low-strength steel is predicted to generally yield at 110,000 lb (489 kN). Depending upon the safety factor used to ascertain that these failure loads are sufficiently higher than the required weight P, it would appear then that the high-strength steel is the preferred choice. However, it is important to note here that the predictions suggest that the high-strength steel would fracture well before general yielding, which implies catastrophic failure without warning; moreover, the actual fracture load would be markedly dependent on the presence of defects in the steel. If this were a safety-critical application, involving potential loss of life, e.g., with people walking under the load, then a far safer option would be to choose the lower strength material; although its load-carrying capacity is somewhat less, it would show obvious signs of yielding long before it fractured. Indeed, virtually all structural alloys, used for pressure vessels and piping, for transportation or in the construction industry, are not ultrahigh strength; most are lower strength with tensile strengths typically below 100 ksi (~700 MPa) to avoid the obvious danger of fracture without warning.

One final point here is whether these predicted failure loads are conservative or not in estimating failure, as it clearly makes engineering sense to be somewhat conservative. Here, the yielding loads would be fairly accurate by assuming failure when the far-field stress exceeds the tensile strength, although we have been more conservative by taking the yield strength instead. With regards to fracture though, there is the issue of small-scale yielding (SSY) and plane-strain conditions, which we have tacitly assumed by using a K_{Ic} value. As described in Section 3.3.2, whether this is the case depends on the size of the plastic zone r_y compared to the uncracked ligament size b and sample thickness B. Using Eq. (3.4), the value of r_y at fracture is roughly 0.01 in. (0.25 mm) in the high-strength steel and the order of 0.5 in. (12.7 mm) in the lower-strength steel. Clearly for the high-strength steel, r_y is small compared to b and B, meaning that SSY/plane-strain conditions apply and the use of K_{Ic} is an appropriate value of the stress intensity at fracture. However, for the low-strength steel, r_y is 0.5 in. and thus no longer at least an order of magnitude smaller than b and B. We are clearly not in plane strain at fracture in this material and so, from Fig. 3.8, the stress intensity at fracture will almost certainly be larger than K_{Ic}, implying that the load at fracture in the mild steel will certainly exceed our predicted load of 320,000 lb (1423 kN), i.e., our original estimate in Table 8.1 for this steel is conservative. This should be enough but as the condition of small-scale yielding is also not realized here, if safety was paramount in this application, it might make sense to additionally do a nonlinear-elastic J-analysis on the fracture loads (as per Section 4.3) in the low-strength steel just to obtain an improved estimate of the fracture load in this material.

8.2.2 Leak-before-break concept for pressure vessels

A safe design criterion for pressure vessels is the "leak-before-break" criterion, i.e., the vessel will leak first, be it gas or a fluid, to give warning of a serious defect before the vessel fails catastrophically. Let us consider the thin-walled pressure vessel, with radius r and wall thickness t (where $r/t \gg 1$), shown in Fig. 8.2, which contains an elliptical surface crack, of depth a and surface length $2c$, aligned along the longitudinal z-direction. In fracture mechanics terms, the leak-before-break concept means that the

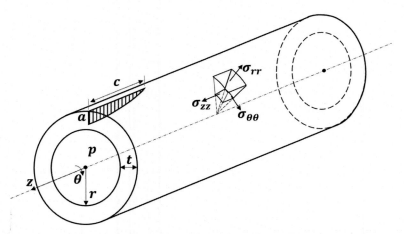

FIGURE 8.2   Schematic illustration of a thin-walled pressure vessel, with radius r and wall thickness t, containing an internal pressure p. The vessel wall contains an elliptical surface crack, of depth a and surface length $2c$, aligned along the longitudinal z-direction. Applying the leak-before-break criterion implies that the critical crack size a_c for fracture must exceed the wall thickness, i.e., $a_c > t$.

critical crack size (for a through-the-vessel-wall crack) has to be greater than the wall thickness, i.e., $a_c > t$.

The stress-state for such a thin-walled pressure vessel containing an internal pressure p is given by the well-known solutions:

$$\sigma_{\theta\theta} = \frac{pr}{t}, \quad \sigma_{zz} = \frac{pr}{2t}, \quad \sigma_{rr} \sim 0. \tag{8.5}$$

The elliptical surface crack in Fig. 8.2 is aligned longitudinally and so is only subjected to the hoop stress $\sigma_{\theta\theta}$.[1] Thus, assuming that linear-elastic fracture mechanics applies, i.e., small-scale yielding in mode I, from Fig. 3.3 in Box 3.2, the K_I solution will be given by:

$$K_I = \frac{1.12 \, \sigma_{\theta\theta} \, \sqrt{\pi a}}{\Phi}, \tag{8.6}$$

where Φ is a function of c/a. Assuming for simplicity that the crack is semi-circular, i.e., $a \sim c$, $\Phi = 1.571$. Accordingly, at fracture when $K_I = K_{Ic}$:

$$K_I = K_{Ic} = \frac{1.12 \, (pr/t) \, \sqrt{\pi a_c}}{1.571}, \tag{8.7}$$

such that the critical crack size, a_c, from Eq. (3.6), is given by:

$$a_c = \frac{1}{\pi} \left(\frac{t \, K_{Ic}}{0.71 \, pr} \right)^2. \tag{8.8}$$

Applying the leak-before-break concept of $a_c > t$;

$$\frac{1}{\pi} \left(\frac{t \, K_{Ic}}{0.71 \, pr} \right)^2 > t, \tag{8.9}$$

then sets the minimum fracture toughness, for the material to make the pressure vessel out of, in order to satisfy leak-before-break:

$$K_{Ic} > \frac{0.71 \, \sqrt{\pi} \, pr}{\sqrt{t}}. \tag{8.10}$$

---

[1] Had this crack been oriented differently, it would represent a biaxially stressed configuration and amenable to the inclined crack analysis described in Section 3.6.1. To calculate a driving force under such mixed-mode loading, ideally one would incorporate both the mode I and mode II contributions, as in Eq. (3.14), but this would require information on both the mode I and mode II fracture toughness, which are not always readily available (especially in mode II). Under such circumstances, as most cracks invariably propagate in mode I, one can generally obtain a reasonable answer, particularly in higher-strength, lower-toughness materials, by considering just the maximum principal stress normal to the crack surface.

8.2.3 Failure of pre-tensioned bolts for a pressure chamber

The third problem involves pre-tensioned bolts which are torqued to a specific level to leave pre-tension in the bolt, which is typically between one half to three-quarters of its tensile strength. This particular problem involved a pressure chamber for a plastic extrusion press, which was held together using 2 ft. (0.6 m)-long high-strength steel bolts with a nominal diameter of 1−5/8 in. (41.3 mm). The bolts were threaded at each end, as shown in Fig. 8.3 and torqued to give a pre-tension in the bolt of approximately half the yield strength of the steel. They were manufactured from a 0.3 wt.% C alloy steel and heat treated by austenitizing at 900°C, quenched in oil, and then double tempered at 500°C to give a martensitic microstructure. Measured properties of the heat-treated alloy steel are listed in Table 8.2.

During manufacture, within minutes to hours after torqueing the bolts, but before the chamber was even pressurized, several of the bolts started to fracture catastrophically in the threads - by cleavage fracture. This was a huge surprise to the extrusion press company who had made numerous

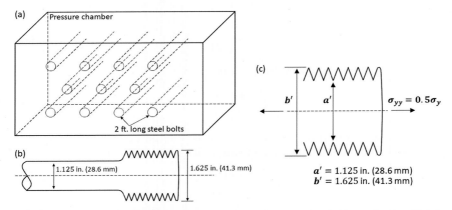

FIGURE 8.3 (a) Schematic illustration of the pressure chamber held together by (b) 2 ft. long, 1-5/8 in. diameter threaded bolts; (c) shows the idealization of the thread as a circumferentially notched rod loaded at a tensile stress of half the yield strength of the alloy steel.

TABLE 8.2 Strength and toughness properties of the quenched and 500°C double-tempered alloy steel (in both English and SI units).

Yield strength, σ_y	Tensile strength (UTS)	Charpy V-notch energy	Fracture toughness, K_{Ic}
60 ksi (420 MPa)	80 ksi (550 MPa)	18 ft.lb. (24.5 J)	70 ksi√in. (77 MPa√m)

such pressure chambers, but never one this big with such large diameter bolts. This was the problem posed to one of the current authors - why were the bolts fracturing and how could this problem be solved?

A previous consultant had suggested that hydrogen embrittlement may have been involved. As described in Section 7.2, hydrogen can most certainly dramatically embrittle steels and markedly degrade the stress intensity to initiate fracture (Table 7.1), but it was not at all clear where the hydrogen would have come from. A far more likely source was the thermal shock generated at the threads of the large bolts during quenching from 900°C, which could certainly generate tiny cracks at the base of the threads.

To assess the likelihood of fracture in the threads, the bolts can be modeled as a circumferentially notched rod in mode I tension, subjected longitudinally to a tensile stress of half the yield strength, i.e., $\sigma_{yy} \sim 30$ ksi (207 MPa), as shown in Fig. 8.3c. At the root of the notch, i.e., in the threads, the K_I solution is given in the Tada, Paris and Irwin handbook [1], in terms of the tensile load P and the dimensions a' and b', as:

$$K_I = \frac{\sqrt{2}P}{\pi a'^2} \sqrt{\pi a' (1 - a'/b')} \cdot Y , \qquad (8.11a)$$

where the geometry factor Y is given by:

$$Y = 1 + 0.5 \left(a'/b'\right) + 0.375 \left(a'/b'\right)^2 - 0.363 \left(a'/b'\right)^3 + 0.731 \left(a'/b'\right)^4. \qquad (8.11b)$$

Substituting the dimensions of the bolts, $a' = 1.125$ in. (28.6 mm), $b' = 1.625$ in. (41.3 mm), and the applied tensile stress of 30 ksi (207 MPa), gives a value of the stress intensity K_I at the root of the threads after torqueing of ~70 ksi$\sqrt{}$in. (77 MPa$\sqrt{}$m), which is essentially the same as the fracture toughness K_{Ic}! Although the size of the bolts do not satisfy the small-scale yielding and plane strain criteria to fully justify the use of K_{Ic}, the closeness of this value to the likely stress intensity developed in the threads was a major cause for concern.

Clearly the use of this alloy steel here with this level of pre-tension was a poor choice of material; it simply had inadequate toughness for this application. A better suggestion would have been to use a AISI 4340 steel (composition in wt.% of 0.4C, 2Ni, 1Cr, 0.2Mo, bal. Fe) or its Si-modified 300-M variant, which can be quenched and tempered to ~100 ksi (690 MPa) strength level but with a ~40% higher fracture toughness. Specifying an "aircraft quality" composition, which has lower S and P impurity content (<0.04 wt.%), would further lower the ductile-to-brittle transition temperature and improve fracture resistance. However, if concerns about hydrogen uptake proved to be real, using a higher-alloyed steel, which would be more resistant to environmentally-assisted cracking, would be prudent. This would naturally increase the cost, but a 200-grade maraging steel (composition in wt.% of 0.03C, 18Ni, 8Co, 5Mo, 0.5Ti, bal. Fe) here would be a good choice because at this strength level, maraging steels have a much higher toughness and are far more resistant to hydrogen embrittlement.

An interesting addendum to this incident was a somewhat similar event that occurred in 2013 near the completion of the building of the new eastern section of the San Francisco-Oakland Bay Bridge. To prevent the bridge separating from the bridge supports during severe earthquake activity, the bridge was designed with so-called shear keys, which are large concrete blocks beneath the road way at the bridge columns supported by large diameter bolts that serve to dampen any seismic oscillations. These bolts were especially large - roughly 2—4 in. (~50—100 mm) in diameter - and were manufactured from quenched and tempered ASTM A354 Grade BD steel, heat-treated to a tensile strength of 140 ksi (965 MPa) and then further hot-dip galvanized for corrosion protection.

The anchor rods were installed into the shear keys in November 2008, but not tensioned until four-and-a-half years later. They were pre-tensioned to 70% of their specified tensile strength, but after a mere 2 weeks one-third of the anchor bolts had fractured, all at or near the threaded section of the bolts. Once the level of pretension was lowered to ~40%, the fractures stopped, but the entire bolted structure had to be abandoned (it could not be removed) and a replacement anchoring system installed as a surrounding "collar"; this occurred only 6 months before the bridge was to be opened.

Subsequent investigation revealed intergranular cracks emanating from the threads of the fractured bolts, which are a characteristic indication of hydrogen-induced cracking in these steels. This hydrogen embrittlement is thought to have occurred because the pre-tensioned bolts were immersed in sea water - an oxidizing environment and stress are prime ingredients to permit hydrogen ingress into the steel [2]. As with the pressure chamber example described above, the thermal heat treatment, involving quenching to martensite prior to tempering, would further likely have led to quench cracks at the root of some of the threads, which could only exacerbate the problem.

One of the current authors attempted to estimate the stress intensity K_I developed at the root of the threads in these large bolts for a pretension of 100 ksi (i.e., 690 MPa, 70% of the tensile strength of the steel), using Eq. (8.11), and found K_I values of ~90 ksi$\sqrt{in}$. (99 MPa$\sqrt{m}$). These values are very high and would be within 10% to 20% of the fracture toughness of this steel at this strength level. Accordingly, they represent a potential safety problem even without further degradation from thermally induced quench cracks or hydrogen embrittlement. Thus, it is little wonder that so many of these bolts fractured after being pre-tensioned. It is just somewhat alarming that with so many advances in the development of materials and the analysis of failures over the past 50 years or so that this can occur with a conventional steel in the 21st century. Our own opinion is that insufficient attention was paid to a fracture mechanics analysis, which should have *red-flagged* this situation long before it happened - but of course, as always, "hindsight is 20/20."

8.2.4 Estimation of the safe life of heart valve prostheses

In the 1970s, mechanical heart valve prostheses, which were medically implanted to replace damaged or malfunctioning aortic or mitral valves in the heart, were primarily manufactured in the form of a metal ring with a tilting disk (occluder) hinged between inlet and outlet struts welded, or integral to, the ring to allow the disk to open and close. The most popular valve employed a Haynes 25, termed L605, alloy (composition in wt.% of 50Co, 20Cr, 15W, 10Ni, 3Fe, 1.5Mn) with a pyrolytic carbon-coated graphite (or pure pyrolytic carbon) occluder. However, *ca.* 1979 one of the most widely implanted valves, the Björk-Shiley "convexo-concave" (C-C) valve, started to fail in vivo. Specifically, fatigue fractures occurred at the base of the outlet strut in the region of the weld, leading to the escape of the occluder. The C-C valve was withdrawn from the market in 1986 but not before over 600 of the 80,000 implanted valves had failed, with two-thirds of those patients dying.

To our knowledge, no life-prediction analysis had been performed on this valve design prior to it entering service, and so at the time one of the current authors was asked to develop a fracture control plan and to present a life-prediction strategy to minimize such in vivo failures. Prior to that time, medical devices were designed on traditional stress-life (*S-N*) methodologies, if at all, where estimates of the cyclic *in-service* stresses were compared to the material's fatigue endurance strength defined at an appropriate number of cycles. However, as such procedures cannot account for the presence of pre-existing cracks and defects, this called for a fracture mechanics approach where the life is computed in terms of the number of cycles for the largest undetected crack to grow to failure, although such a methodology had been rarely, if ever, utilized in the medical device industry previously. The analysis was performed on a similar valve, the Shiley Monostrut valve (Fig. 8.4), where the inflow and outflow strut were integral with the metallic ring, i.e., no welds were involved [3].

As shown in Fig. 8.4b, c, the operation of the occluder in tilting-disk valves puts maximum bending stress at the base of the struts (where the welds were located on the C-C valve). To perform a conservative analysis for the Monostrut valve, we (i) considered minimum dimensions, but within specifications, of the valve components, (ii) used finite-element analysis to calculate the stresses at the base of the struts, which relied on experimental pulse duplicator studies to determine the maximum total vertical load values from measured deflections on calibrated struts under simulated peak physiological flow conditions, (iii) analyzed the effect of realistic pre-existing flaws located in these peak stress locations, and (iv) finally measured fatigue-crack propagation rates as a function of ΔK in the identical Haynes 25 alloy used to make the valves, over a range of R-ratios in 37°C Ringer's lactate solution to simulate a physiological environment (Fig. 8.5).

For the inlet strut under peak physiological loading, the maximum computed tensile stress ranges were indeed found to act at the base of the strut, at the vertex furthest away from the outlet strut (at location V in Fig. 8.4a) during valve closing; corresponding numerical analysis for the outlet strut similarly revealed that maximum tensile stress ranges act near the base on the outflow (downstream) surface although now during valve opening [3]. These stress ranges were found to be $\Delta\sigma^{\infty} = 34$ MPa for the outlet strut and 76 MPa for the inlet strut. Consideration should also be made of any residual stresses existing at the base of the struts; this was not

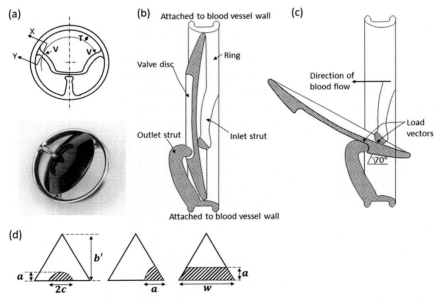

FIGURE 8.4 (a) Björk-Shiley Monostrut prosthetic heart valve, showing schematic diagrams of the geometry of the inlet and outlet struts, and (b, c) the operation of the occluder in the opening and closing of the valve, and (d) possible crack configurations in either strut. *Adapted from Ref. R.O. Ritchie, P. Lubock, Fatigue life estimation procedures for the endurance of a cardiac valve prosthesis: stress/life and damage-tolerant analyses, J. Biomech. Engin. Trans. ASME 108 (1986) 153—160 and reproduced by permission.*

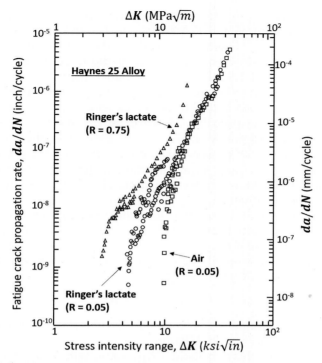

FIGURE 8.5 Variation in fatigue-crack propagation rates (da/dN) as a function of stress-intensity range (ΔK) for Haynes alloy 25 tested at a load ratio of $R = 0.05$ and 0.75 in air and Ringer's lactate solution at $37°C$ [3]. *Reproduced by permission.*

done in the original analysis but is invariably an important factor to account for.

Various idealized crack configurations were analyzed (Fig. 8.4d). In particular, worst-case corner and through-thickness cracks, of depth a, were considered at the base of both the inlet and outlet struts (Fig. 8.4a-c). Numerical solutions for these configurations gave the mode I K_I-solutions, in terms of the tensile stresses, σ^∞, at the base of the struts under peak physiological loading, to be:

$$K_I = Y\sigma^\infty \sqrt{\pi a} , \qquad (8.12a)$$

where for a corner crack, $Y = 0.8$ for $a/b' \leq 0.25$, and 0.95 for $a/b' = 0.50$, and for the through-thickness crack, for $0 < a/b' < 1$:

$$Y = \sqrt{\frac{2b'}{\pi a} \tan\left(\frac{\pi a}{2b'}\right)} \left[\frac{0.923 + 0.199\left(1 - \sin\frac{\pi a}{2b'}\right)^4}{\cos\left(\frac{\pi a}{2b'}\right)} \right] . \qquad (8.12b)$$

For initial crack size of 25% of the strut dimensions, which was within the validity range of all the K_I-solutions in Eq. (8.12), using the peak physiological stresses determined by the pulse duplicator and finite-element studies, the stress intensity K_I levels did not exceed 3.3 MPa√m in the inlet strut and 1.5 MPa√m in the outlet strut. In terms of non-destructive evaluation (NDE) procedures for crack detection, the relevant dimension on crack size is the surface crack length, with penny-shaped cracks being the most probable. However, stress intensities developed ahead of such cracks are typically at least 20% lower than for corner cracks. Accordingly, for conservatism, corner cracks were taken as the optimum solution for both struts in terms of detectable surface crack lengths; through-thickness cracks were deemed to be highly unlikely and thus overly conservative.

In light of these small stress intensities, the pertinent variation in da/dN with ΔK was estimated quantitatively by taking a regression fit to the near-threshold fatigue-crack growth data in Fig. 8.5, specifically below 10^{-7} mm/cycle in the simulated physiological environment with nominal zero-tension loading ($R \sim 0$) loading, viz.:

$$da/dN = (7.10 \times 10^{-20})\Delta K^{12.2}, \qquad (8.13)$$

for units of MPa√m and m/cycle, where the threshold stress intensity range is given by $\Delta K_{TH} \approx 4.5$ MPa√m.

A comparison of this threshold ΔK_{TH} value with the computations based on Eq. (8.12) indicated that the initial ΔK values developed under peak physiological loading for initial surface crack sizes, a_o, less than half of the cross-sectional dimension of both the inlet and outlet struts were less than ΔK_{TH}. Thus, provided appropriate NDE procedures are enacted and can reliably detect all incipient cracks in excess of a surface length of roughly 800 μm for this valve, the initial ΔK values at the base of the struts under peak physiological loading will be below the threshold such that fatigue-crack propagation leading to device failure during a patient's lifetime would be unlikely.

A more conservative approach is to disregard the concept of a threshold and to assume that fatigue can continue at very low ΔK values to vanishingly small growth rates. As described in Section 7.5, using such a damage-tolerant design procedure (Eqs. 7.19–7.22), a finite life can be predicted by integrating the relevant fatigue-crack growth relationship, in this case Eq. (8.13) (after substituting in K_I-solution in Eq. 8.12), between the limits of the assumed initial crack size, a_o, and crack size at final failure, a_f, which can be defined by the critical crack size a_c (Eq. 3.6). In terms of the relevant Paris law constants - m (= 12.2) and C (= 7. 10 × 10⁻²⁰ in units of MPa√m and m/cycle) - and the peak physiologically induced stress range, $\Delta\sigma^{\infty}$, calculated for each strut, this gives the number of cycles to cause failure N_f as:

$$\int_{a_o}^{a_f} \frac{da}{a^{m/2}} = C\,(Y\Delta\sigma^{\infty})^m\,\pi^{m/2} \int_o^{N_f} dN . \qquad (8.14)$$

For the calculated peak physiological stresses developed at the base of the inlet and outlet struts, respectively, Eq. (8.14) can be evaluated to give the estimated fatigue lifetime N_f as a function of initial flaw size a_o, which can be interpreted as the largest undetected crack size. In lieu of the fracture toughness, failure was defined conservatively as a surface crack size of ~50% of the cross-sectional dimension of either strut. Projected lifetimes are plotted in Fig. 8.6 assuming ~38 × 10^6 cycles (heart beats) per year. It is clear that the life of the valve is a strong function of the resolution of the prior NDE inspection technique, i.e., of the largest undetected flaw size.

Based on these computations, the damage-tolerant analysis predicts that, for continuous peak physiological loading, and peak stresses of 0–76 MPa at the base of the inlet strut and 0–34 MPa at the base of the outlet strut, it should take in excess of 3 × 10^9 cycle, i.e., approximately 90 years, to grow a fatigue crack to cause failure of either strut, *provided all initial cracks, of surface lengths exceeding 500 μm, are detected by NDE prior to the device entering service.*

For the interested reader, the full analysis is detailed in Ref. [3] and considers other secondary factors such as residual stresses and the role of small cracks (as outlined in Section 7.4.3), and the corresponding stress-life analysis. The prime utility of the fracture mechanics analysis is that it sets a quantitative target (in terms of pre-existing defect size) for the non-destructive evaluation of the valves which can be related through Fig. 8.6 to a conservative estimate of the lifetime. By considering worst-case assumptions of physiological conditions, peak stresses at critical locations, and a zero life for crack initiation, and coupling this with a sound NDE of the device, one can realistically conclude that this particular Monostrut valve, unlike the welded C-C value, would be unlikely to fail in service.

There is also an interesting addendum to this incident as the fall-out from these failures was so intense that Shiley Inc., which had been bought in 1979 by Pfizer Inc., ceased to manufacture mechanical heart valves, and so they never actually marketed the Monostrut valve. Moreover, as implanting surgeons at that time became concerned about the structural durability of such metal valves, this led to the rising popularity of heart valves made of a ceramic-like material, specifically pyrolytic carbon, which is a turbostratic carbon material [4] originally used as re-entry heat shields for space vehicles. Like the tilting disk in the

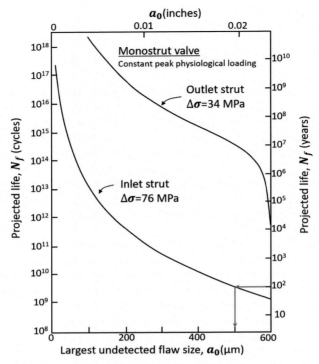

FIGURE 8.6 Predicted fatigue lifetimes as a function of the pre-existing defect size, a_o, for the peak stresses in the inlet and outlet struts of the Monostrut valve. Since the heart beats at roughly 38×10^6 cycles per year, for a device lifetime exceeding patient lifetimes, conservatively taken as 100 years, a maximum detectable defect size can be estimated (as shown by the *red arrows*), which is the maximum size of pre-existing defect that must be detected by quality control procedures prior to the sale of the device to minimize the prospect of the device failing in vivo [3]. *Reproduced by permission.*

metallic valves, the pyrolytic carbon was either used as a coating on graphite or as a monolithic material. The St. Jude Medical bi-leaflet valve, which was made as a ring of the coated material with two matching semi-circular leaflets press-fit into ring which opened and closed to regulate blood flow, was one particularly successful prosthetic device of this type. The corresponding damage-tolerant lifetime analysis for this valve [5] followed the same procedures outlined above for the metal valve, but was far more complicated because pyrolytic carbon is a prototypical brittle material. In contrast to the metallic Haynes 25 alloy which has a fracture toughness exceeding ~60 MPa√m, the toughness of pyrolytic carbon is between 1 and 2 MPa√m. Additionally, as shown in Fig. 7.16a, its Paris law exponent m can be as high as 50, which is far in excess of any metallic material, meaning that the predicted lifetimes become highly sensitive to the stress (raised to the 50th power!), with a factor of two lower (yet still very high) sensitivity to crack size (as per Eq. 8.14). The difficulties of utilizing damage-tolerant life-prediction strategies for brittle materials are described in Section 7.5 - for this reason, ceramics are rarely used in safety-critical applications - but were successful for this particular heart valve prosthesis by using caution and conservatism at every stage of the analysis. The interested reader is referred to Ref. [5] for further details.

References

[1] H. Tada, P.C. Paris, G.R. Irwin, The Stress Analysis of Cracks Handbook, third ed., ASME, New York, NY, 2000.

[2] Y. Chung, L.K. Fulton, Y. Chung, Environmental hydrogen embrittlement of G41400 and G43400 steel bolting in atmospheric versus immersion services, J. Fail. Anal. Prev. 17 (2017) 330.

[3] R.O. Ritchie, P. Lubock, Fatigue life estimation procedures for the endurance of a cardiac valve prosthesis: stress/life and damage-tolerant analyses, J. Biomech. Engin. Trans. 108 (1986) 153–160. ASME.

[4] R.H. Dauskardt, R.O. Ritchie, Pyrolytic carbon coatings, in: L.L. Hench, J. Wilson (Eds.), An Introduction to Bioceramics, World Scientific Publ. Co., Singapore, 1993, p. 261.

[5] R.O. Ritchie, Fatigue and fracture of pyrolytic carbon: a damage-tolerant approach to structural integrity and life prediction in "ceramic" heart-valve prostheses, J. Heart Valve Dis. 5 (Suppl. 1) (1996) S9–S31.

Suggestions for further reading

Once you have mastered the topics in this primer, there are several textbooks available for further reading. These are listed below with our personal commentary.

1 T.L. Anderson, Fracture Mechanics: Fundamentals and Applications, 4th ed., CRC Press, Boca Raton, FL, 2017.

This is by far the most comprehensive, readable and up-to-date advanced textbook on fracture mechanics available.

2 J.F. Knott, Fundamentals of Fracture Mechanics, Butterworths, London, UK, 1973.

The first book on fracture mechanics, now rather dated, but an interesting early perspective on the subject, including an introduction to some of the basic mechanics of elasticity theory.

3 D. Broek, Elementary Engineering Fracture Mechanics, 4th ed., Kluwer, New York, NY, 1991.

Another interesting, but relatively elementary and dated, treatment of the subject.

4 S.T. Rolfe, J.M. Barsom, Fracture and Fatigue Control in Structures, 2nd ed., Prentice-Hall, 1987.

Also rather dated, but nevertheless this is a highly useful text for the practicing engineer.

5 A. Saxena, Advanced Fracture Mechanics and Structural Integrity, CRC Press, Boca Raton, FL, 2019.

A modern advanced treatment of nonlinear elastic fracture mechanics, and especially of creep crack growth.

6 B.R. Lawn, Fracture of Brittle Solids, 2nd ed., Cambridge Univ. Press, 1993.

A sound description of fracture mechanics specifically for ceramics.

7 S. Suresh, Fatigue of Materials, 2nd ed., Cambridge Univ. Press, 1998.

The most comprehensive treatment on fatigue and fatigue crack growth.

Index

Note: 'Page numbers followed by "f" indicate figures and "t" indicate tables.'

Printed in the United States
by Baker & Taylor Publisher Services